Brick Stone Metal Wood
BUILDING ON TRADITION

BRICK has been a mainstay in architecture from ancient times through to the modern age. Flexible and resilient, it continues to be used with great ingenuity in contemporary building projects.

Brick Stone Metal Wood

BUILDING ON TRADITION

images
Publishing

Published in Australia in 2020 by
The Images Publishing Group Pty Ltd
ABN 89 059 734 431

Offices

Melbourne
6 Bastow Place
Mulgrave, Victoria 3170
Australia
Tel: +61 3 9561 5544

New York
6 West 18th Street 4B
New York, NY 10011
United States
Tel: +1 212 645 1111

Shanghai
6F, Building C, 838 Guangji Road
Hongkou District, Shanghai 200434
China
Tel: +86 021 31260822

books@imagespublishing.com
www.imagespublishing.com

Copyright © The Images Publishing Group Pty Ltd 2020
The Images Publishing Group Reference Number: 1526

All rights reserved. Apart from any fair dealing for the purposes of private study, research, criticism or review as permitted under the Copyright Act, no part of this publication may be reproduced, stored in a retrieval system or transmitted in any form by any means, electronic, mechanical, photocopying, recording or otherwise, without the written permission of the publisher.

 A catalogue record for this book is available from the National Library of Australia

Title: Brick Stone Metal Wood: Building on Tradition
Author: Carlos García Fernández and Begoña de Abajo Castrillo [Introduction]
ISBN: 9781864708370

Printed on 140gsm Gold East matt paper by Everbest Printing Investment Limited, in Hong Kong/China

IMAGES has included on its website a page for special notices in relation to this and its other publications. Please visit www.imagespublishing.com

Every effort has been made to trace the original source of copyright material contained in this book. The publishers would be pleased to hear from copyright holders to rectify any errors or omissions.

The information and illustrations in this publication have been prepared and supplied by the contributors. While all reasonable efforts have been made to ensure accuracy, the publishers do not, under any circumstances, accept responsibility for errors, omissions and representations, express or implied.

Contents

6 From Material Samples to Tectonic Actions

BRICK

14 **Weaving Together the Strands of History** United Kingdom
22 **Coming Together** Indonesia
28 **Green and Open Space** Vietnam
36 **Letting the Light In** Colombia
46 **Brick in the Wall** United States
56 **Private Spaces** Brazil
64 **Reclaiming Acoustics** Poland
72 **Historical Tradition** France

STONE

80 **Echoing the Ocean** Portugal
88 **Curtain Wall** China
96 **Luxury in Stone** Spain
104 **Elegant Lake Views** Switzerland
112 **Pre-Hispanic Splendor** Mexico

METAL

122 **Looking at the Heroic Age** Turkey
130 **Blending In** France
140 **Revisiting Ancient Tradition** China
146 **A-Framing the View** United States

WOOD

156 **Private Access** Poland
162 **New Façades** Spain
172 **The Heart in Ikast** Denmark
182 **Stacking the Terraces** Thailand
188 **Rise Like a Phoenix** Switzerland
194 **Treasure Box** Israel
202 **Student Rhythm** United States
208 **Disappearing Trick** Norway
216 **Planned Spontaneity** Spain
226 **Communal Education** United Kingdom
236 **Making It** Italy
244 **Worthwhile Delays** Chile

254 Project Credits

From Material Samples to Tectonic Actions

This book surveys a large number of projects, sweeping across continents and ranging from small residential buildings to big institutional complexes, and forms a collection of works that aims to represent the state of the art in the use of traditional materials in contemporary architecture. The projects, organized by the primary materials used, are examples of how revisited traditional techniques or new technologies are employed to create spaces and effects by means of tectonics and its expressive potential in architectural design.

The aforementioned organization of the works based on materials seems appropriate in order to classify such a diverse and rich selection of projects, echoing all those materials that traditionally have been present in architecture and construction. Needless to say, raw materials are normally transformed and processed by the industry before their arrival to construction sites, and it is interesting to understand how technology is changing the way in which architects today are using materials in their works, shifting from traditional or vernacular handcraftship to a global and more industrialized offering.

To understand the ideas that emerge from the selection of projects, we are going to follow a process of disassembly, decomposing the buildings into smaller components, or samples, which are not always exclusive of a particular material. Then we are going to identify the actions used by the architects to bring every element together to create larger entities that conform to the shape or the spatial definition of their works. The resulting repository of elements and actions will help to understand how choices of materials are part of the design process in order to achieve a specific effect based on needs or desires.

The creation of collections or repositories has been a recurrent strategy in the work of artists (and architects) throughout history, and two clear examples of this are Carl André's *chart of available forms* and Richard Serra's *"Verb List."*

On one hand, the collection of available forms by Carl André[1] shows a series of elements organized in two tables, with the names clearly labelled and accompanied by an illustration. It is a list of the forms that the artist frequently used, matching the shapes of different industrial elements found on the market after being manufactured or processed. In such a way, the artist separates the elements so they can be autonomous from one another and are available to be incorporated into his work.

On the other hand, Richard Serra's verb list[2] compiles a series of actions whose objects are the hypothetical contextual conditions, the elements, or the unnamed material. This collection of verbs would serve him as a basis to work across different mediums and contexts, as well as with diverse materials.

In both examples, the lists compiled are carefully crafted so the elements they contain can be used with varying degrees of freedom, favoring their instrumental and open character. In the case of André, the list includes elements without referring to the actions that could be used on them whereas Serra's list contains verbs that refer to actions without a specific object to be applied to.

These lists show an evident intent by the artists to gather fragments of existing reality, independently of their materiality, and then actively incorporate them into their projects. In these, the information compiled is given a format, and through this an argument is created for their works; a manifesto that prioritizes

---

[1] The charts of available forms were selected by the artist to illustrate the invitation card for his 1984 exhibition at the Konrad Fischer Galerie in Düsseldorf.

[2] "Verb List Compilation: Actions to Relate to Oneself" (1967–68), first published in 1971 in the journal *Avalanche*, and now part of MoMA's collection.

action over the object in the case of Serra, and that values the object over action in the case of André.

Going through the selection of works included in this book it is possible to identify "available forms" or material samples and "verbs" or tectonic actions that bring those elements together by means of organization strategies in the pursuit of an effect or architectural statement.

In this context, to organize means to arrange something systematically or similarly, managing and working with the elements we have available to create an order with a certain intention. The concepts of arranging and systematizing relates either to the design process or the final built form. From a historical perspective, the classic order uses composition as the tool to create something from a number of things, placing them in a particular manner. On the other hand, the modern order refers to position, understood as the precise location of something with regard to a frame of reference. The equivalent in the contemporary order is arrangement, the "preparation" in contrast to the single, unmovable position of those parts that define a whole. "To arrange," as opposed to the idea of designing through composition, implies a systematic and analytical thought process. The result of the arrangement, as opposed to unifying and timeless design, understands architecture as a problem of elements and relationships, assembly strategies, or organization.

In architectural projects we work with multiple organizations that lead us to do more than compose, but to fragment. And so, as in any organizing process, before organizing it is necessary to previously define and understand the parts or material samples that we are going to work with.

This technique can be recognized in the work of contemporary architects, not only as something inherent in an architecture that can be decomposed into architectural elements, but also as a tool to design from the fragmentation and identification of discrete elements.

When working with a set of things or material samples, we tend to group or classify them, establishing a specific arrangement and therefore a certain relationship between them, which entails a new condition of relative positions with regard to the rest of the elements.

To position or place implies establishing relationships. Here, what is important is knowing, on the one hand, how to define the criteria to position one part with regard to the rest, and on the other hand, how to identify the parameters or concepts on which to base and establish the general criterion of the organization.

Therefore, the projects can be analyzed to extract conclusions about how to operate with organization systems of the materials based on parts and the rules that bring them together, depending on different criteria, to build a major whole. The aim of this classification is to identify the design processes behind these systems allowing a better understanding of the endless possibilities of using traditional materials nowadays.

First of all, we have tried to identify how the different materials have been transformed in order to create a list of units or minimum elements similar to Carl Andre´s list. The resulting list extracted from the projects collected in the book could be endless, but we have chosen what we feel are the main ones: "bricks," "sticks," "beams," "slabs," or "panels" among others.

The way in which the elements are implemented during construction could be defined by a tectonic action. This action depends at the same time on the form of the unit, but also on the result or effect that the architect wants to achieve. We have also defined a list of actions that can be combined with the pieces counted in the first list: '"to separate," "to offset," "to pile," "to eliminate," "to overlap," or "to perforate" are some of them.

Finally, we could also analyze the effects generated by the properties of the piece and the action used for its implementation. Some of these effects are: "texture," "shadow," "irregularity," "movement," "continuity," or "filter" among many others.

The following organization shows how different materials can be transformed in similar pieces or construction components, by using each project as an example that illustrates the parts, actions, and effects relationship explained above.

BRICKS

Many of the projects use the element "brick" as the minimum material sample in their design. Bricks normally have their own entity and self-supporting qualities, which allows them to be arranged by means of gravity in a bottom-up logic. They can be piled using cement or adhesives and it is the way they relate to the adjacent ones that creates the different effects in the space.

In Apartment in Binhn Thanh the combination of different terra cotta elements provides a texturized façade that controls sun shading and natural ventilation with pieces designed to allow light and ventilation through its shape. Every single element is attached to the ones next to it separating the sets of the same type by horizontal structural lines that frame the organization of the pieces and create different light and texture effects on the façade

Something similar happens in KS Residence but in this case the identical ceramic bricks are arranged according to the level of privacy desired or the needs of climatic control and natural ventilation. The translucent and homogeneous effect of the ceramic lattice happens thanks to an equidistant displacement of the bricks.

A little displacement in the bricks that compose the façade is also the action that leads the design in the Warwick Street project where a colored and varnished ceramic piece is repeated throughout the façade creating a pattern based on the regular displacement of every piece in relation with the adjacent ones and therefore a vibrant and texturized effect in the vertical columns of the façade.

In House for Solidarity, the texturized aspect of the façade is designed and controlled during construction using BIM software. The pattern is created by the rotation of the bricks in seven different positions using rear templates to create a weaving effect on the façade that adds texture by means of light and shadow effects on the ceramic wall.

Another method using bricks can be seen employed in Fundación Santa Fe. In this case with a different proportion of its dimensions. The tensile system of steel rods holds the ceramic pieces creating a permeable filter that covers all the façades shifting the logic of gravity arrangements that exist in all the previous examples.

It is important to remember that brick as an available form is not exclusive of ceramics materials but there are also some examples in natural stone. One of these examples show the different ways of working with stone today.

In Huayacán the construction took the form of piling volumetric stone elements together. The walls are built using a composite system of concrete and volcanic stones that compose a continuous and monolithic surface made of irregular pieces. For the construction, local labor was used in a direct application of traditional techniques combined with contemporary materials and modernist lines.

STICKS

The "stick" is present in several projects in the book, mainly in the wood section due to the predominant use of this form in the construction of lattice screens that wrap the exterior of the buildings. This element has one of its dimensions much longer than the others, which makes it a form that needs the existence of a secondary structure to link all of them when creating surfaces and controls the offset of the different pieces to create diverse effects as can be seen in the examples in the book.

In Hjertet the wooden lattice is present both inside and outside, creating a nice relationship between the building and the surroundings as well as generating warm interiors. The wood sticks have diverse behavior in different climatic conditions acquiring patina when exposed to natural climate conditions.

One example of engineered façades using sticks as main construction elements is Ibiza Gran Hotel. Here, the veil that covers the façade is made out of wooden and resin composite sticks with an aluminum core, an application of technology to achieve a better performance of the construction element that also allows longer dimensions of the piece. It creates an extremely light appearance that blurs the presence of the openings behind the screen.

A last example of the works using the stick as primary unit is Inter Crop Office building where the louvers covering the glazed façades are a reinterpretation of the traditional wooden screens using an artificial element made out of aluminum that fakes the wooden aspect of its finishing.

PLANKS

The "plank" is a planar element with a negligible thickness and variable proportions in the surface, still having one dimension predominant in relation to the others. We find examples that use strips in wood and metal and the fixation techniques are similar independently of the material itself.

The first example using this type of form in wood is Arkansas Bear Claw, where the cedar screens cover the façade unifying the volume through a single material. Behind the screens, the openings are organized to control the views and to optimize the energy efficiency of the envelope. The different types of windows create a variety of situations, filtering the light through the wooden lattice.

Other examples include buildings that are clad in wood creating continuous and mainly opaque surfaces. This happens in Ogden Centre for Fundamental Physics at Durham University, where the disposition of the cladded timber larch elements in a diagonal layout emphasizes the dynamic composition of the volume.

Similarly, in Steinhardt Museum of Natural History, the faceted wooden volume resembles the keel of a boat or a geological formation. The wooden planks clad the concrete façade and cover the insulation layer, producing a vibrant effect due to the subtle variations in color and finishing of the different elements that compose the whole surface.

Holmen Aquatics is a highly energy efficient building that uses larch wooden planks to envelope the volume and to create subtle shadows behind the dark screens. Wood is used in this case to mimic the context during summer and to contrast with the snow landscape in wintertime.

The dark façades in Polyvalent Hall combine a metallic envelope and a series of wooden lattices that follow a repetitive vertical rhythm that echoes both materials through its shape. On the inside, cladded fir tree panels build a warm space that drives the views toward the landscape.

SLABS

Different to the "planks," the "slabs" are planar elements with a considerable thickness. They are mainly used to clad other supports and to render the surfaces with a certain quality in relation with the context.

In Wenzhou Century Park Culture Club the stone pieces are organized using a curtain wall technology that creates a smooth and continuous surface. Every piece has the same dimension and they are cut according to the folded shape of the building volume.

Meanwhile, in Bürgenstock Hotel the limestone slabs are cladded to the façade creating smooth surfaces that contrast with the natural bedrock on which the building rest. It is an artificial and processed interpretation of the same natural material existing on the building site.

Something similar occurs in Akelarre Hotel where the stone slabs cover both the external and internal walls building a rationalized rock volume from which the openings seem to be carved out. The materiality of the walls provides a compact and monolithic appearance to the building in relation with the surrounding landscape.

As a last example of this type of element and in a more urban context is Atlantic Pavilion where stone slabs are combined with exposed concrete elements, seeking to find a balance between the natural material and the artificial one.

PANELS

Some examples use 'panels' as primary elements and those can be made out of different materials but also using diverse techniques in its construction. The logic that drives the assembly of panels is based in prefabrication. The panel, which could be technically decomposed in smaller fragments, works in term of design and construction as an irreducible entity.

Two cases on which the use of metallic panels is predominant are Shui Cultural Center and the New Entrance Hall of Museum of the Middle Ages. In the first example the bronze sheets cover the entire façade, which is also a pitched roof that allows the light to enter the space. The subtle reflections on the surfaces create a double lighting effect during the day and night, when the building projects light from the inside. In the second one the aluminum panels have a different dimension, reliefs and perforations. This combination of finishing using the same material offers a vibrant appearance of the façade that changes with the light and the different moments of the day.

A different technique of building prefabricated panels is by combining bricks as happens in Kent State Center for Architecture and Environmental Design. The bricks are arranged in a fashion that by means of vertical frames the pattern is misaligned, producing a fragmentation in the façade and affecting the perception of the scale and the dynamism of the building surfaces.

BEAMS

The 'beam' as a design element is used in a set of projects featured in the book. Even though the dimensions of this type of element are bigger than the other categories, they follow similar logics of assembly and organization to create bigger sets.

Examples of interlaced systems of beams are found in Berluti Manufacture: Factory and Development Center and House behind the Roof. In the factory, the main structural elements are interlaced to span over the main space of the building, covering a large diaphanous nave. In the house, the sloped roof is built with glulam timber that is both exposed on the outside and the inside of the house.

Similarly to the two previous examples, in IMPLUVIUM, the main glulam structure that covers the inner plaza was fabricated using CNC technology and assembled on site. The vertical structure also combines wood and steel to create composite columns that holds the roof from the outside, letting the public space extend to the interior in the ground floor.

A more complex structural system is seen in Gimnasio Municipal de Salamanca where different wood structures are combined in order to control direct sun exposure and control the views. In this case the structure is organized in order to deceive the view and create the illusion of a weightless materiality of wood and metal surfaces.

RECLAIMED FRAGMENTS

"Reclaimed fragments" as primary elements for construction are not as frequently used as the other categories but they have a great potential following the logics of recycle and up-cycling processes. As their name implies, these are materials that have been recovered from previous use, and brought back into use as new construction elements.

The most relevant example in the book using this type of element is CKK Jordanki. The reclaimed ceramic brick fragments are embedded on the concrete slabs and walls, and becoming partially detached after the wall is cured. This innovative use of the ceramic brick provides better acoustics compared with smooth concrete walls, but also bestows an attractive and textured appearance to the building.

Summing up, the understanding of the parts that configure a whole, and the relationships that are established among them and with the whole, is crucial for the design process. If a fundamental part of designing consists in assembling or organizing these parts to configure a new whole, the identification and previous classification of said parts is also part of this process, with the possibility of this organizing process becoming a project in itself in order to create a desired effect.

The definition of every project finds equilibrium between many parameters that need to be taken into account during the design process. Those parameters are related to the choice of the material and the decision of the form in which it will be implemented. Both material and form depend on several intrinsic conditions such as its technical restrictions, the availability of the material, the proximity of the needed factory that transforms it, etc. There are also extrinsic conditions that vary the results like the programmatic and contextual aspects of every project, the cultural implications, the materiality of the surroundings, or the requirement of light among others. Last but not least, there are also subjective intentions that lead the design process in order to reach the aimed results.

It is interesting to see how different materials can be transformed into similar units, or how different materials can be implemented to generate similar effects. From this reflection, we might wonder if it is possible to reach similar needs with different materials.

Each project in this book finds a balance between the identification of the material samples and the overall effect produced by means of tectonic actions. Both technical decisions, such as anchoring systems, manufacturing processes, and material properties, or aesthetics qualities—such as color, texture, size, or aspect—come into play when deciding to select one material over the others.

Many of these projects reflect how the industry is offering architects endless possibilities to design with traditional materials, transforming them or implementing them in new ways that perpetuate their use nowadays.

Begoña de Abajo and Carlos García

Begoña de Abajo Castrillo

Begoña de Abajo Castrillo holds a Masters from Columbia University in New York where she studied as a Fulbright Fellow, and graduated with honors from the Madrid School of Architecture (ETSAM) obtaining the School End of Studies Prize. She also studied at the Illinois Institute of Technology in Chicago, and was a member of the Foster + Partners team in Madrid. She is currently a PhD researcher at the Architectural Projects Department at ETSAM, and is a principal and co-founder of the architecture practice RAW/deAbajoGarcia.

Carlos García Fernández

Carlos García Fernández holds a PhD in Architecture and a Masters from Columbia University in New York where he studied as a Fulbright Fellow. He has studied at the Madrid School of Architecture (ETSAM) and also at TU Delft in The Netherlands. As an independent architect he has won several prizes in architectural ideas competitions by himself and working with other colleagues. He has been a scholar at the Spanish Academy in Rome and a researcher at Keio University in Tokyo. He is currently working as an Associate Professor at the Architectural Projects Department at ETSAM. He is a principal and co-founder of the architecture practice RAW/deAbajoGarcia.

www.deabajogarcia.com

Weaving Together the Strands of History

WARWICK STREET
COMMERCIAL
BRICK, STONE, GLAZED TILES
LONDON, UNITED KINGDOM
2018

Squire and Partners has extensively remodelled an existing 1980s building to create a vibrant development that maintains its previous office and retail uses, but makes a more meaningful connection with the surrounding Soho Conservation Area by creating a contemporary crafted design. Designed for the Royal London Mutual Insurance Society, the architects exploited the layers of history of the building in order to remodel and reimagine, rather than demolish and replace, offering more sustainable and time-saving possibilities that create more from less.

In the case of Warwick Street the full site history was explored, starting with the origins of the site as home to master fabric weavers Holland and Sherry's, through the abstract and rebellious spirit of local Soho resident Francis Bacon, to the postmodern 1980s façade.

Traditional elements of the building, such as the façade, mansard roof, and reception desk evolve through these layers and emerge surreally enhanced, capturing the playful and creative spirit of Soho.

High-quality office and retail spaces are provided within the existing building structure behind a new façade, while a new top floor is expressed as an irregular folded mansard. The striking new façade establishes a rational frame of glazed Umbra Sawtooth bricks with stone spandrels, animated with vertical bands featuring "woven" green tiles that pay tribute to the area's textile history, and relate to the glazed tiles which define the surrounding streets. A series of full height glazed bays at ground floor create impactful retail units, while a canopied entrance marks the office reception.

The metal-clad top floor creates a bold interpretation of a modern mansard roof, the erratic playful form making a nod to Soho's history of rebellion and the inherent creativity which defines the local culture.

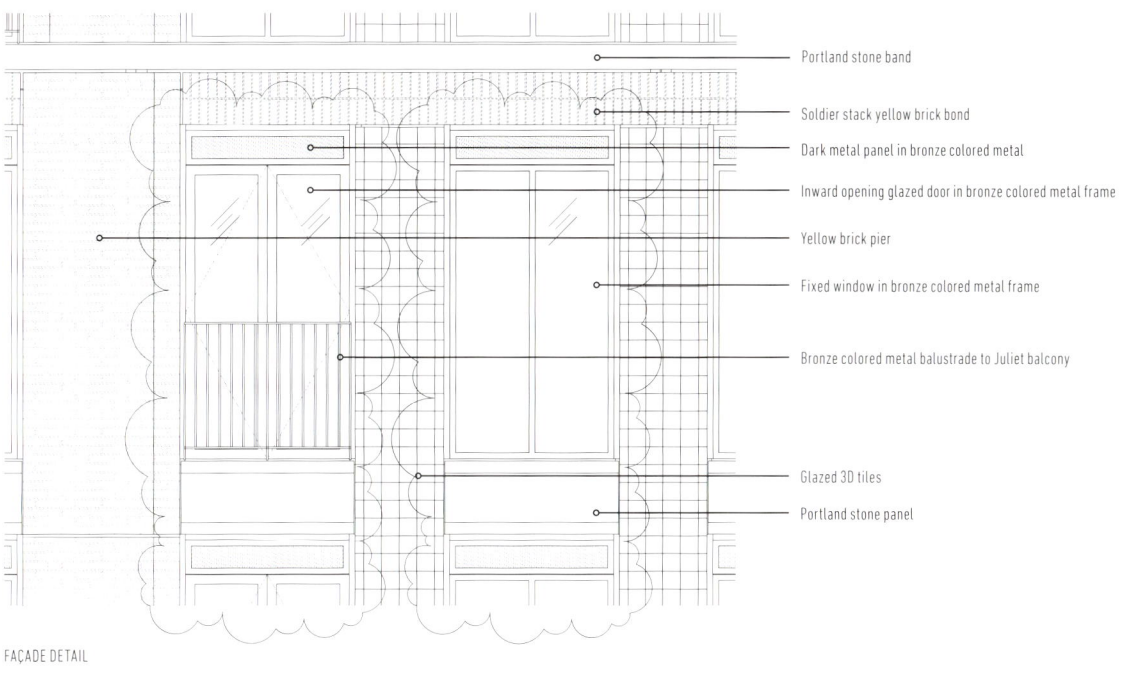

- Portland stone band
- Soldier stack yellow brick bond
- Dark metal panel in bronze colored metal
- Inward opening glazed door in bronze colored metal frame
- Yellow brick pier
- Fixed window in bronze colored metal frame
- Bronze colored metal balustrade to Juliet balcony
- Glazed 3D tiles
- Portland stone panel

FAÇADE DETAIL

Weaving Together the Strands of History

Coming Together

SINGKAWANG CULTURAL CENTER
CULTURAL
BRICK
SINGKAWANG, INDONESIA
2017

This striking design was the culmination of a project that sought to revitalize an inactive cinema space, transforming it into a place of activity and bringing together people from the major ethnic groups for communal activities and events, combining both public space and tourism.

The architects were keen to preserve the spirit of the old site, and used a combination of brick and steel to create a new skin. In addition, the materials used are local Singkawang bricks, giving a nod to the various historical, aesthetic, and quality values.

The local brick contains an iron oxide, giving it a distinct rust hue in pink-orange-red, and kaolinite, giving it a soft white tone. The kaolin also produces the unique characteristic style of the resulting product: robust and distinctive.

Type A brick arrangements

Type B brick arrangements

Void

Steel capping

Corten steel signage with lights installed behind

Steel column

Steel plate around the window list

FAÇADE ELEVATION

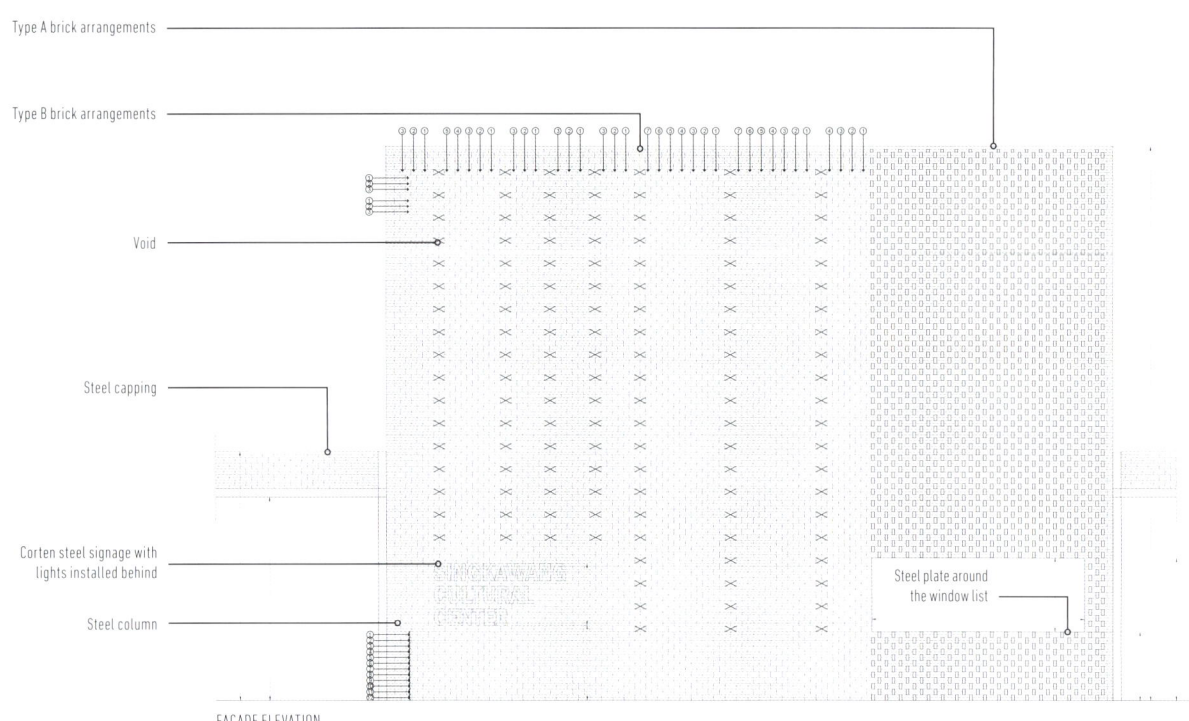

26 Brick Stone Metal Wood

Green and Open Space

APARTMENT IN BINH THANH
RESIDENTIAL
BRICK, TERRA COTTA TILES, CONCRETE
HO CHI MINH CITY, VIETNAM
2016

A terra cotta block screen is a familiar and popular method of creating shade and privacy in Vietnam. This small apartment with seven rooms in Binh Thanh district showcases the potential of the brick screening.

The new apartment sits on an unusual and slightly curved shape, the result of merging several plots of different sizes with a total length of around 430 feet (40 meters). The architects prepared a design for an attractive building with lots of open spaces.

At the center of the apartment is a large open courtyard. A large void connects the garden-space on the ground floor to the open-area courtyard. Each room exposed to the central courtyard was given a "lanai" (semi-outdoor space). All these outside spaces are organically connected with each other, allowing fresh air and daylight to circulate through the whole of the building. Access to plentiful outdoor space enhances the occupants' living style, and in addition provides a deep connection to the outdoors. These spaces were generously planted with fast-growing tropical plants, making use of the natural abundance of natural daylight and natural ventilation, and overall creating an enjoyable outdoor space.

Local materials were sourced where possible in order to keep construction costs low. The exterior brick façade screen employs several different patterns of the terra cotta blocks, creating a characteristic front to the building. The screen not only enhances the outdoor space in the courtyard, but also makes use of natural ventilation and passive cooling through shade, while simultaneously enhancing occupant security, and resulting in a truly beautiful exterior.

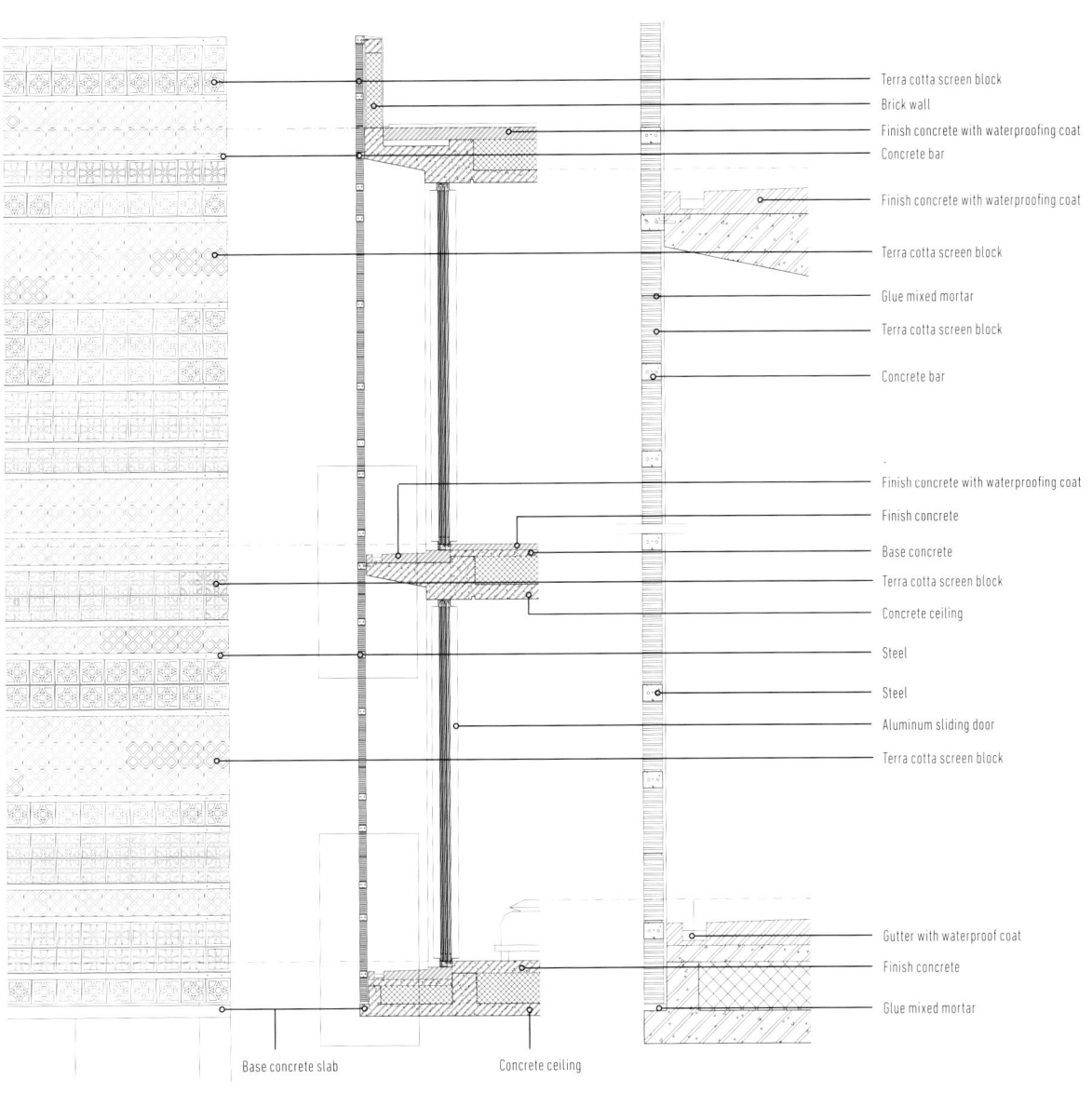

FAÇADE DETAIL

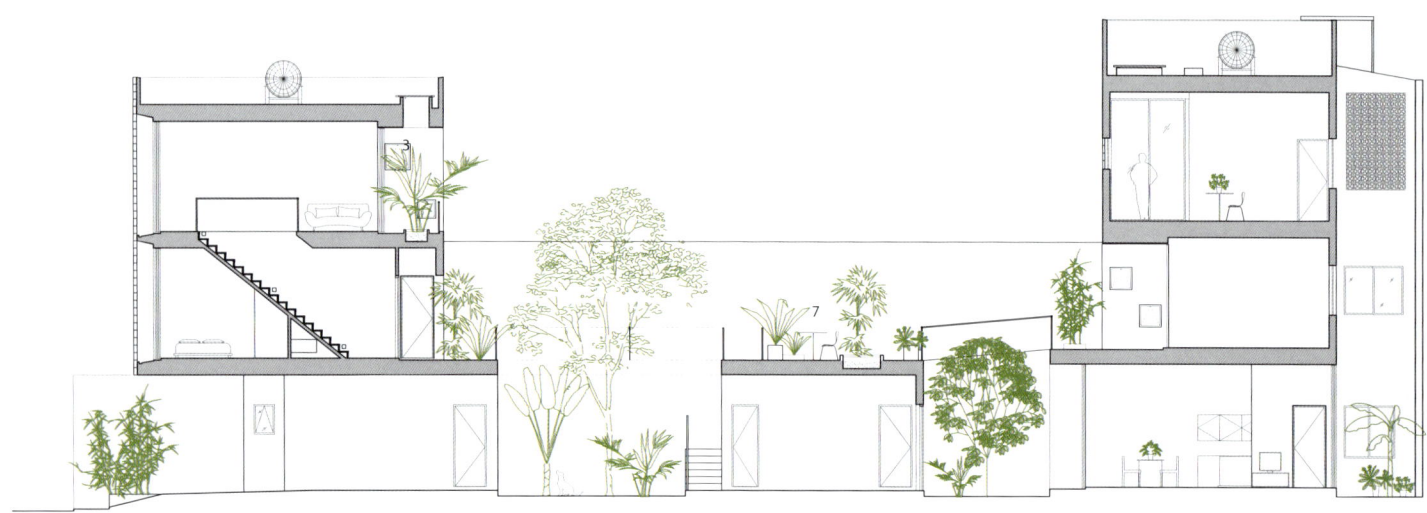
SECTION

34 Brick Stone Metal Wood

Letting the Light In

FUNDACIÓN SANTA FE
MEDICAL
BRICK, STEEL, CONCRETE
BOGOTÁ, COLOMBIA
2016

This stunning brick expansion of the Fundación Sante Fe de Bogotá hospital not only succeeds in creating a connection between the existing hospital buildings and itself, but also culminates in a striking new landmark on the cityscape. The use of brick was a deliberate decision to continue the identity of the current medical complex, but it is the innovative and creative use of the brick that results in the building's character. Comprising 12 levels, the new structure forms an imposing sight, but also allows for further expansion in the future.

The impressive brick façade employs metal parts and cables that support the actual bricks. This allows the architects to employ different patterns and textures, resulting in a variety of different natural lighting and the ability to change it depending on the needs. Externally, the different patterns enhance the overall character of the building, offering variety in the façade appearance.

In line with the Fundación's philosophical principles, the façade provides the focus for helping a patient's recovery through the provision of more natural light. The solarium, in particular, seeks to provide a garden hospital concept, and allows the patient to have greater contact with the external surrounds, reducing stress and helping to aid their overall recovery. Additional filtered light penetrates the interior of the hospital through the brick façade as well as through light wells set within the building.

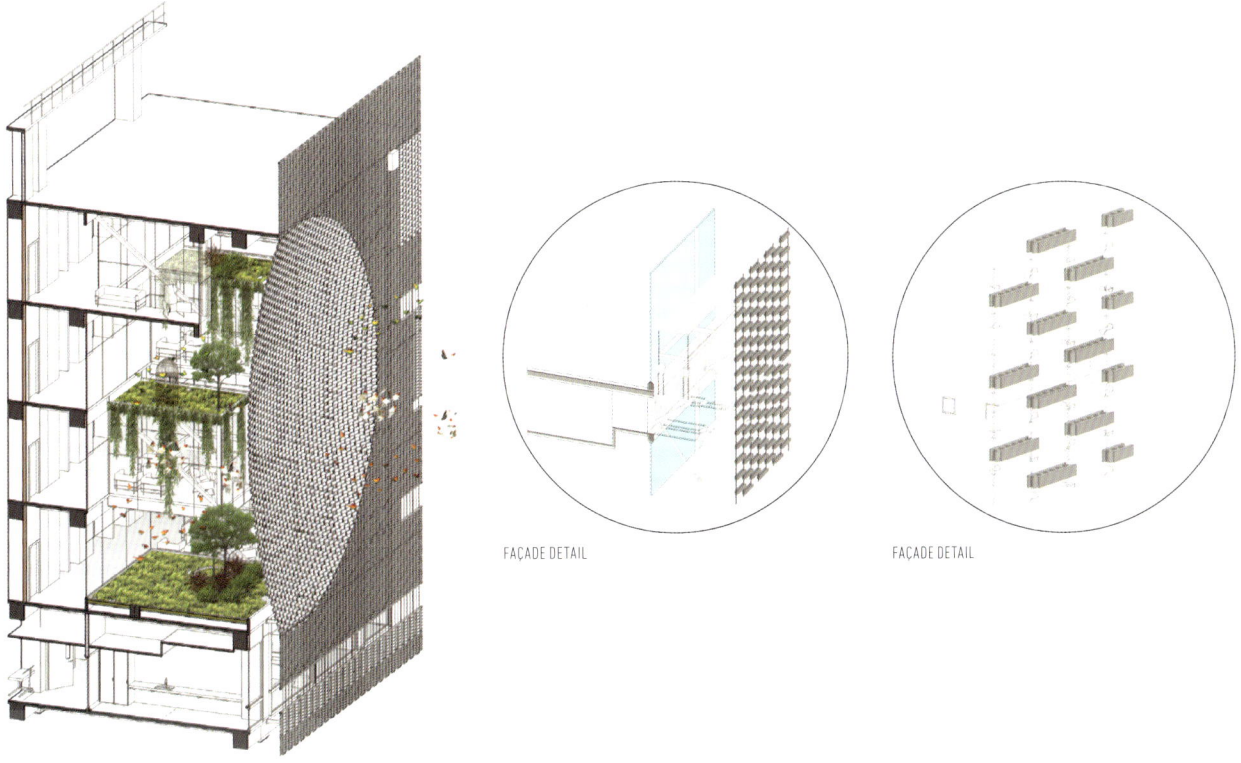

SECTION

FAÇADE DETAIL

FAÇADE DETAIL

Letting the Light In   39

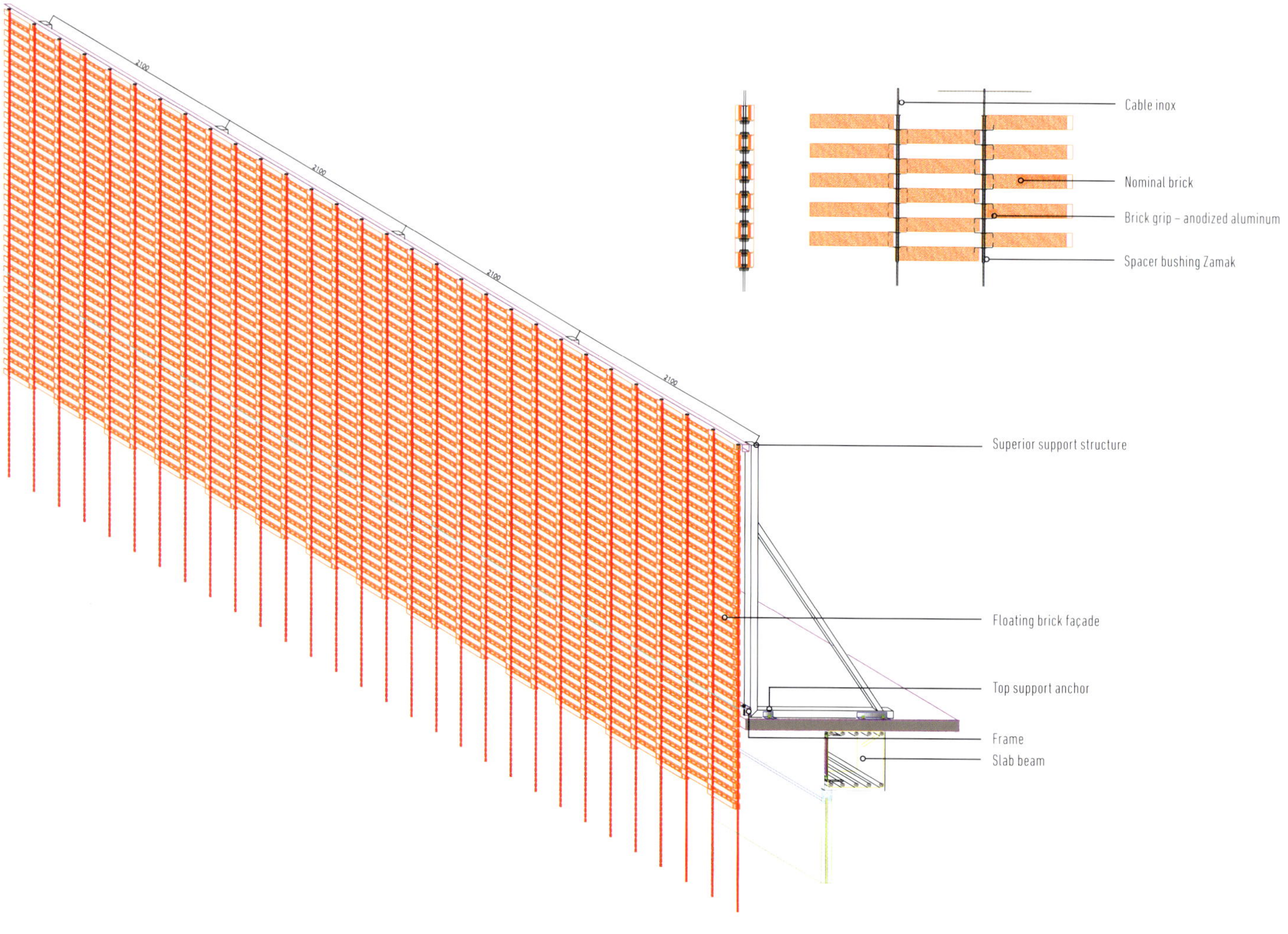

Cable inox

Nominal brick

Brick grip – anodized aluminum

Spacer bushing Zamak

Superior support structure

Floating brick façade

Top support anchor

Frame

Slab beam

Brick Stone Metal Wood

Letting the Light In 43

Brick in the Wall

KENT STATE CENTER FOR ARCHITECTURE AND ENVIRONMENTAL DESIGN
INSTITUTIONAL
BRICK, CONCRETE, GLASS, WOOD
OHIO, UNITED STATES
2016

The Kent State Center for Architecture and Environmental Design "Design Loft," the winner of an international competition, establishes an innovative center for the design disciplines. Sited strategically at a hinge between the campus and city, the Design Loft forms a new hub that forges connections between Kent State University and the recently revitalized downtown Kent.

An expansive studio loft forms the heart of the program. Designed to maximize flexibility, it will accommodate a growing program and evolve to meet the educational needs of the architecture and design fields. The tiered arrangement of studios informs the massing of the building, which bridges the institutional and residential scales of its neighbors.

A continuous gallery anchors the building's main public level and opens up to the university's new esplanade, a pedestrian walkway that establishes a connection between the university and the city as part of a joint redevelopment initiative.

The ascending sequence of ground floor spaces support a broad range of activities including a café, gallery, a 200-seat multipurpose lecture room, classrooms, library, and related reading areas. The color and texture of the iron-spot brick façade and custom brick fins, fired in a bee-hive kiln by the local Belden Brick Company, relate to the materials of the surrounding campus and town.

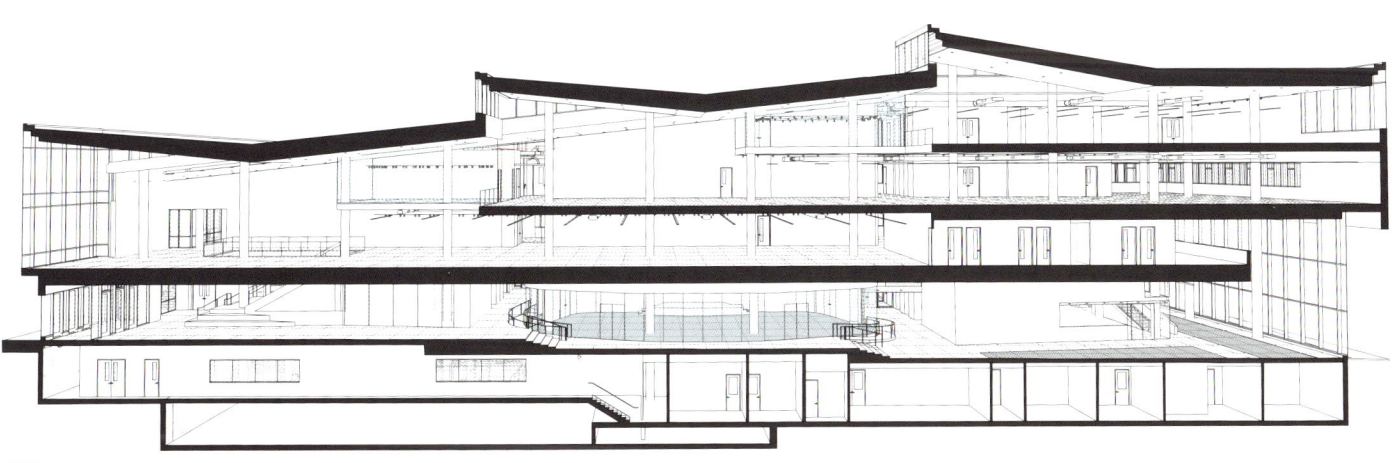
SECTION

Brick in the Wall

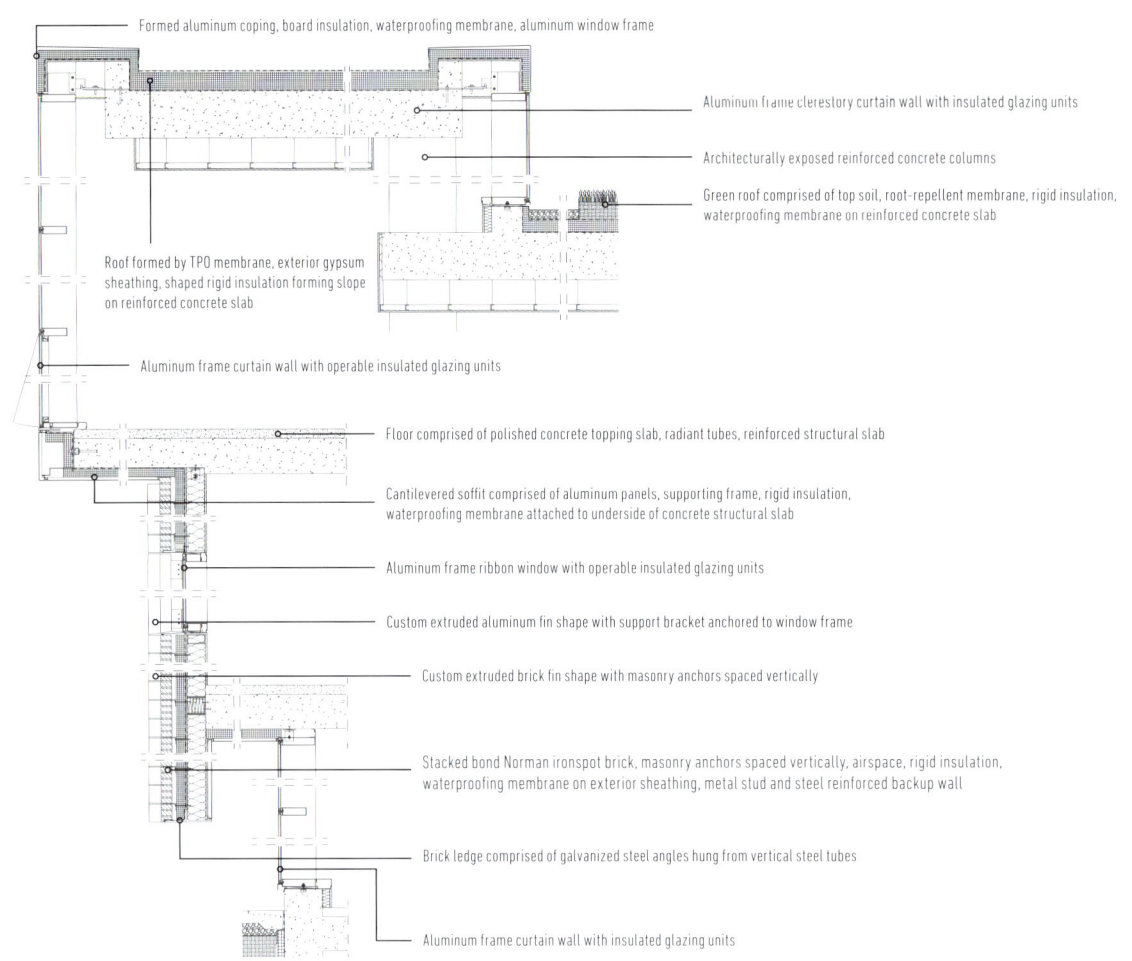

SECTION DETAIL

- Formed aluminum coping, board insulation, waterproofing membrane, aluminum window frame
- Aluminum frame clerestory curtain wall with insulated glazing units
- Architecturally exposed reinforced concrete columns
- Green roof comprised of top soil, root-repellent membrane, rigid insulation, waterproofing membrane on reinforced concrete slab
- Roof formed by TPO membrane, exterior gypsum sheathing, shaped rigid insulation forming slope on reinforced concrete slab
- Aluminum frame curtain wall with operable insulated glazing units
- Floor comprised of polished concrete topping slab, radiant tubes, reinforced structural slab
- Cantilevered soffit comprised of aluminum panels, supporting frame, rigid insulation, waterproofing membrane attached to underside of concrete structural slab
- Aluminum frame ribbon window with operable insulated glazing units
- Custom extruded aluminum fin shape with support bracket anchored to window frame
- Custom extruded brick fin shape with masonry anchors spaced vertically
- Stacked bond Norman ironspot brick, masonry anchors spaced vertically, airspace, rigid insulation, waterproofing membrane on exterior sheathing, metal stud and steel reinforced backup wall
- Brick ledge comprised of galvanized steel angles hung from vertical steel tubes
- Aluminum frame curtain wall with insulated glazing units

SECOND-FLOOR PLAN

GROUND-FLOOR PLAN

1 Studio
2 Faculty suite
3 Advising suite
4 Administration suite
5 Critique room
6 Seminar
7 Lab

Private Spaces

KS RESIDENCE
RESIDENTIAL
BRICK, GLASS, CONCRETE
NATAL, BRAZIL
2016

A desire for privacy guided the design of this residence, closed to the exterior, but open to a large internal space comprising three floors. The architects, working within strict parameters for townhouse construction, sought to maximize these limitations to create as much volume within the building as possible.

The design creates three stories, with each floor design varied rather than simply stacked together. The garage and storage areas are located on the basement level, while the social and living areas are placed at ground level, and the upper levels contain the bedrooms and bathrooms.

The floors rise above the central void, with a variety of ceiling heights and resulting in large air volumes which help to keep the interior cool and aid with air circulation.

Each section of the house is connected via internal walkways from one slab to the other, thereby creating links to each of the residential spaces. The section nearest the street contains the living and entertainment rooms. The utilitarian kitchen, dining room and service areas are located on the ground level, with access to the yard. The upper floor, connected by a footbridge, comprises the private rooms.

The differentiated pagination of the bricks establishes small openings in the façade that allow the illumination and ventilation of the internal spaces. For greater thermal comfort in the upper part of the house a permanent ventilation strip and windows was installed next to the floor, providing a Venturi effect.

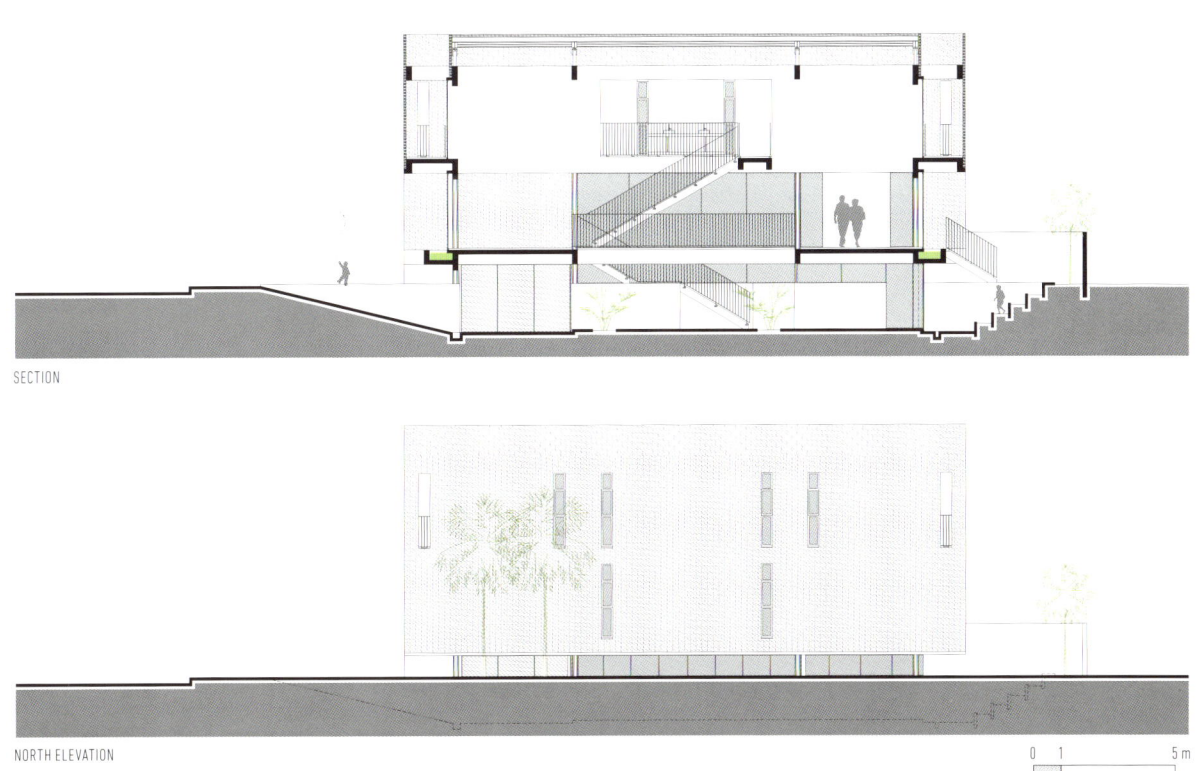

SECTION

NORTH ELEVATION 0 1 5 m

Private Spaces

- Concrete slab
- Concrete beam
- Gypsum plasterboard lining
- Colorless tempered glass window
- Internal and external sill in concrete
- Gravel box for moisture containment
- Waterproofed backing
- Air mattress
- Masonry + smooth plaster + white paint
- Concrete slab
- Natural terrain

TYPICAL DETAIL

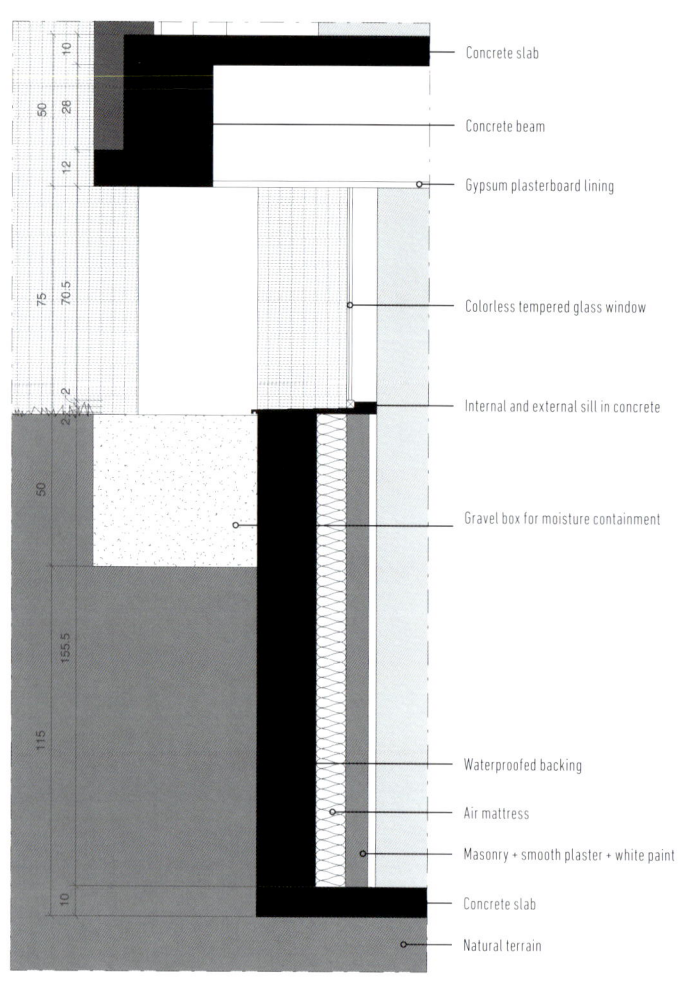

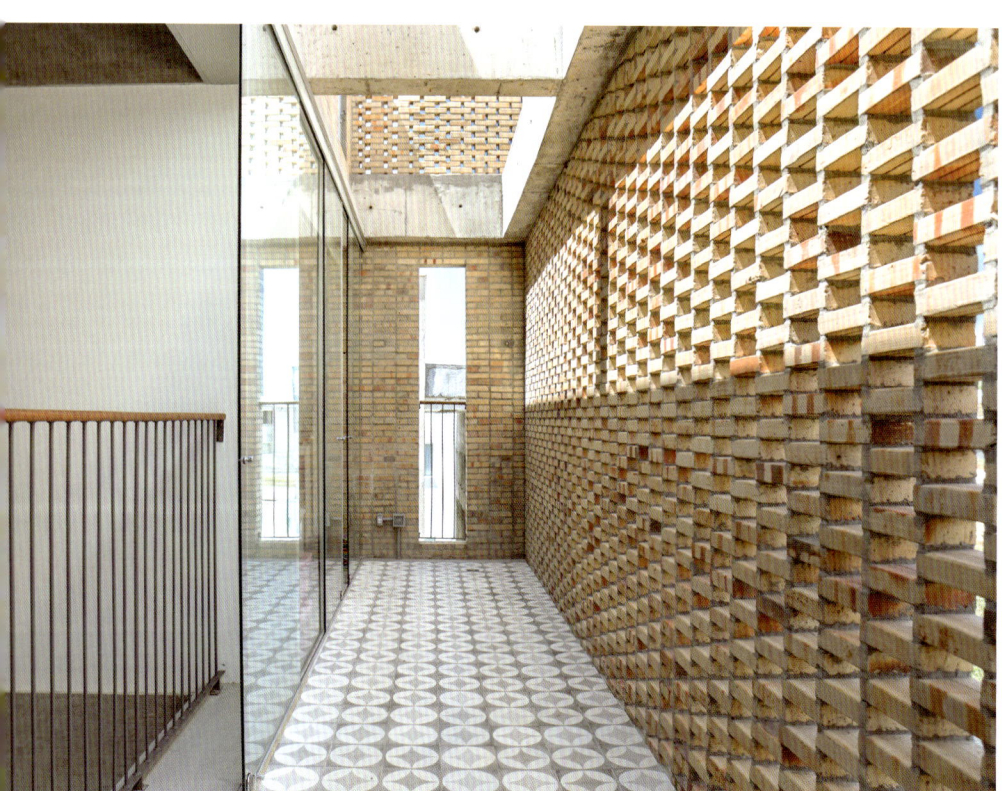

Brick Stone Metal Wood

Private Spaces

Reclaiming Acoustics

CKK JORDANKI
CULTURAL
BRICK, CONCRETE
TORÚN, POLAND
2015

This new concert hall combines a novel building technique but also succeeds in blending a modern structure with existing city structures in the UNESCO-heritage city of Torún. Its compact stature helps it blend with the existing medieval buildings, and the green park provides views to the nearby river.

Architect Fernando Menis devised an innovative technique, picado, which consists of mixing concrete with reclaimed broken-up red bricks. The result is a striking and unusual finish, but with the added bonus of providing excellent acoustics.

The exterior is made of pale concrete. The cutaways in the concrete shell allow glimpses of the red-brick lining, a reference to the traditional brick building façades of the neighboring buildings.

While the outer skin of the concert hall remains rigid, inside the building behaves like a fluid. The program is characterized by great flexibility to the extent that a building which, according to the brief, was meant to be only a concert hall, ended up being a space for all kinds of concerts and events within the same budget.

The auditorium can be easily adapted to different audiences, catering for smaller as well as higher numbers. Thanks to its mobile roof, it can be adjusted to effectively absorb all kinds of performances from symphonic music, chamber music, to theater, opera, and film. Finally, the concert hall can be opened outwards, uniting the interior and exterior space to accommodate outdoor performances or other large events.

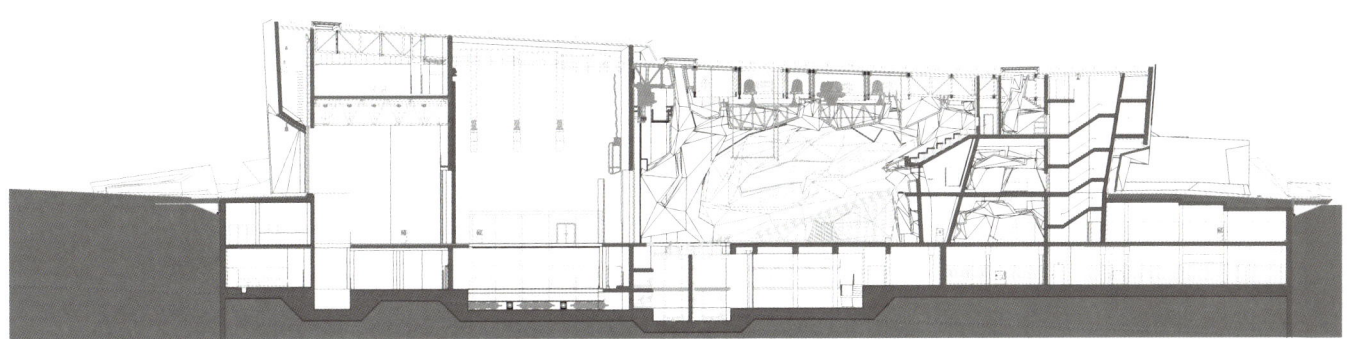
LONGITUDINAL SECTION

Reclaiming Acoustics

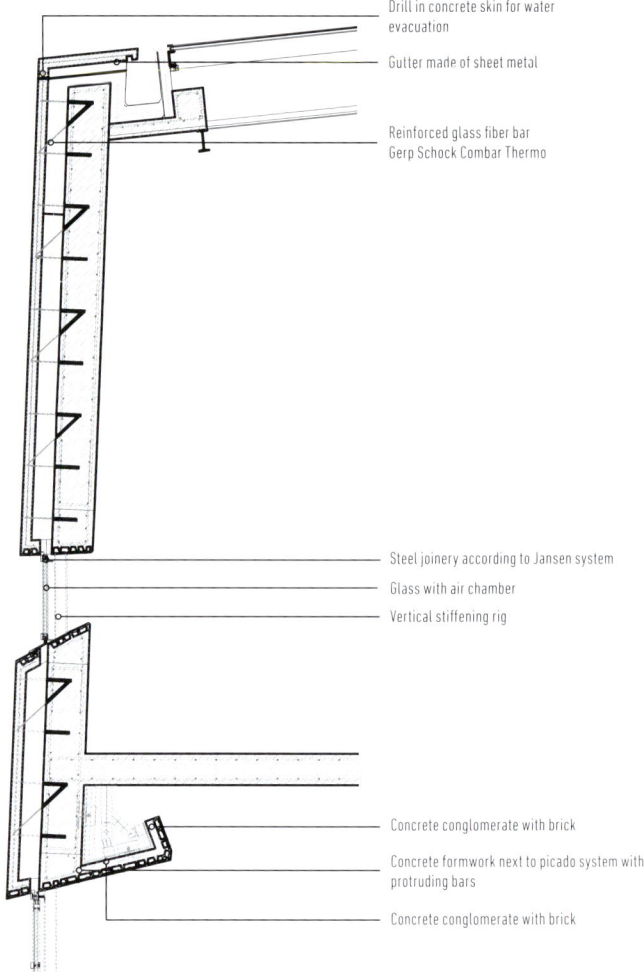

FAÇADE DETAIL

- Drill in concrete skin for water evacuation
- Gutter made of sheet metal
- Reinforced glass fiber bar Gerp Schock Combar Thermo
- Steel joinery according to Jansen system
- Glass with air chamber
- Vertical stiffening rig
- Concrete conglomerate with brick
- Concrete formwork next to picado system with protruding bars
- Concrete conglomerate with brick

70   Brick Stone Metal Wood

Historical Tradition

HOUSE FOR SOLIDARITY
MEDICAL
BRICK, CONCRETE
OISE, FRANCE
2015

House for Solidarity is located on a former military base, a 12-hectare site that has been a wasteland since 1993, and this new construction forms an important part of the renewal of the district. Its elegant silhouette provides a contemporary continuation between the existing buildings and the new structures.

Neither monumental, nor impressive, the project distinguishes itself from nearby housing by the subtle movement of its façades, and its single building material. The use of bricks, a material historically present in Beauvais, and more widely in the north of France, provides a nod to the existing architectural vernacular tradition. The skillful cladding and overlapping bricks forms a weaving pattern of diamonds, an allusion to the tapestry workshop tradition of Beauvais. Shadows and light reveal the pattern, expressing the architectural richness of the project, singular and precise without being ostentatious. Public entry is at the center of the building, while the social workers' offices are located on the first floor.

Using the cavity wall insulation technique, every single brick (about 38,000) of this peculiar patterned façade has been designed individually. The pattern is created by overlapping, overhanging, and rotating each brick in seven different styles. It was designed using a complete 3D model (BIM) with homemade parametric tools. The use of such tools allowed a versatile and adaptable design, facilitating communication with all construction companies and the mounting of the bricks on the construction site.

The House for Solidarity is a key project in this fast-changing district, and the use of brick plays a part in this achievement. A tribute to the diversity of façades in Beauvais, its use expresses a sensitive and welcomed return of this material in today's architecture.

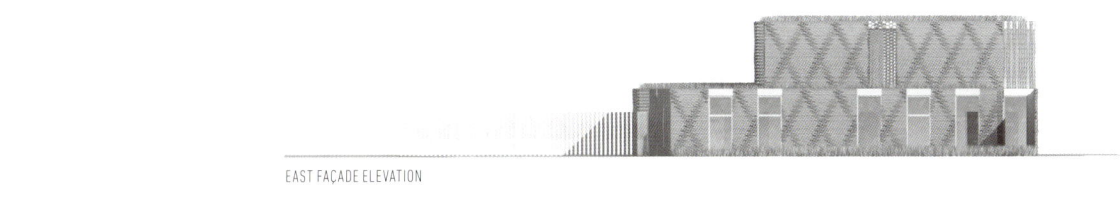
EAST FAÇADE ELEVATION

NORTH FAÇADE ELEVATION

Historical Tradition

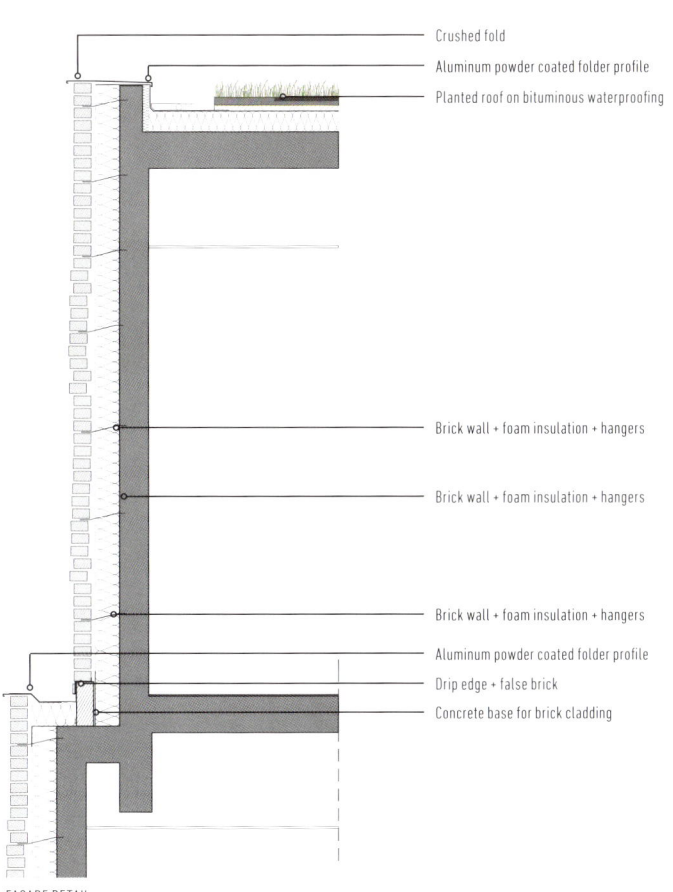

FAÇADE DETAIL

- Crushed fold
- Aluminum powder coated folder profile
- Planted roof on bituminous waterproofing
- Brick wall + foam insulation + hangers
- Brick wall + foam insulation + hangers
- Brick wall + foam insulation + hangers
- Aluminum powder coated folder profile
- Drip edge + false brick
- Concrete base for brick cladding

Historical Tradition 77

STONE has been used in constructing structures since time immemorial. With its myriad tones and textures, it is no wonder that stone is still a popular material and enjoyed by contemporary architects.

Echoing the Ocean

ATLANTIC PAVILION
CULTURAL
STONE, CONCRETE, TILES
VIANA DO CASTELO, PORTUGAL
2018

In the small coastal town of Viana do Castelo in northern Portugal, architect Valdemar Coutinho designed an imposing sports complex, prominently sited on a corner plot. The brief from the local council was for a public complex that would serve the local community. The brief also stipulated tight budget constraints, a challenge which resulted in the building's somewhat brutalist appearance.

The building takes up just about all of the 19,214 square foot (1,785 square meter) site, yet manages to exude an appealing dynamic and a humanized image. Externally, the project is defined by a dramatic structure at the corner of the plot, which cantilevers above the entrance. The main entrance provides room for up to 80 people to wait before entering the arena. A reception area is joined by a cafeteria, restrooms, and a technical support area.

The external cool gray tones of concrete and stone provide a muted palette and reference the nearby ocean. This nautical theme is continued inside the entrance with various panels by artist Mário Rocha of gray embossed tiles, which allude to the crustaceans and algae of the nearby beaches.

- Geotextile blanket
- Layer of rolled and washed pebbles
- Thermal insulation in extruded polystyrene foam boards
- Minimum motar slope
- Ventilated wall performed by stone plates "Azul Cascais" stapled with mechanical attachment, in stainless steel
- Celling acoustic in plasterboard finished with white paint
- Ruffle in natural zinc
- Concrete
- Thermal insulation in extruded polystyrene foam boards
- Thermal insulation in extruded polystyrene foam boards
- Plasterboard finished with white paint

FAÇADE DETAIL

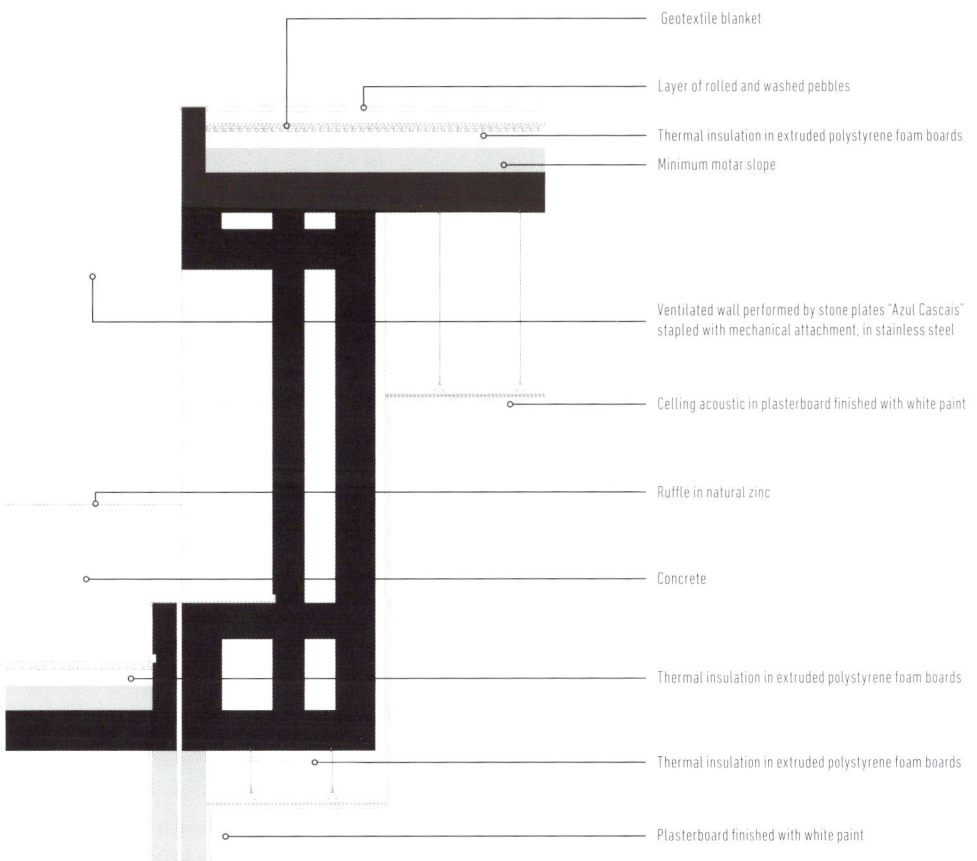

Echoing the Ocean

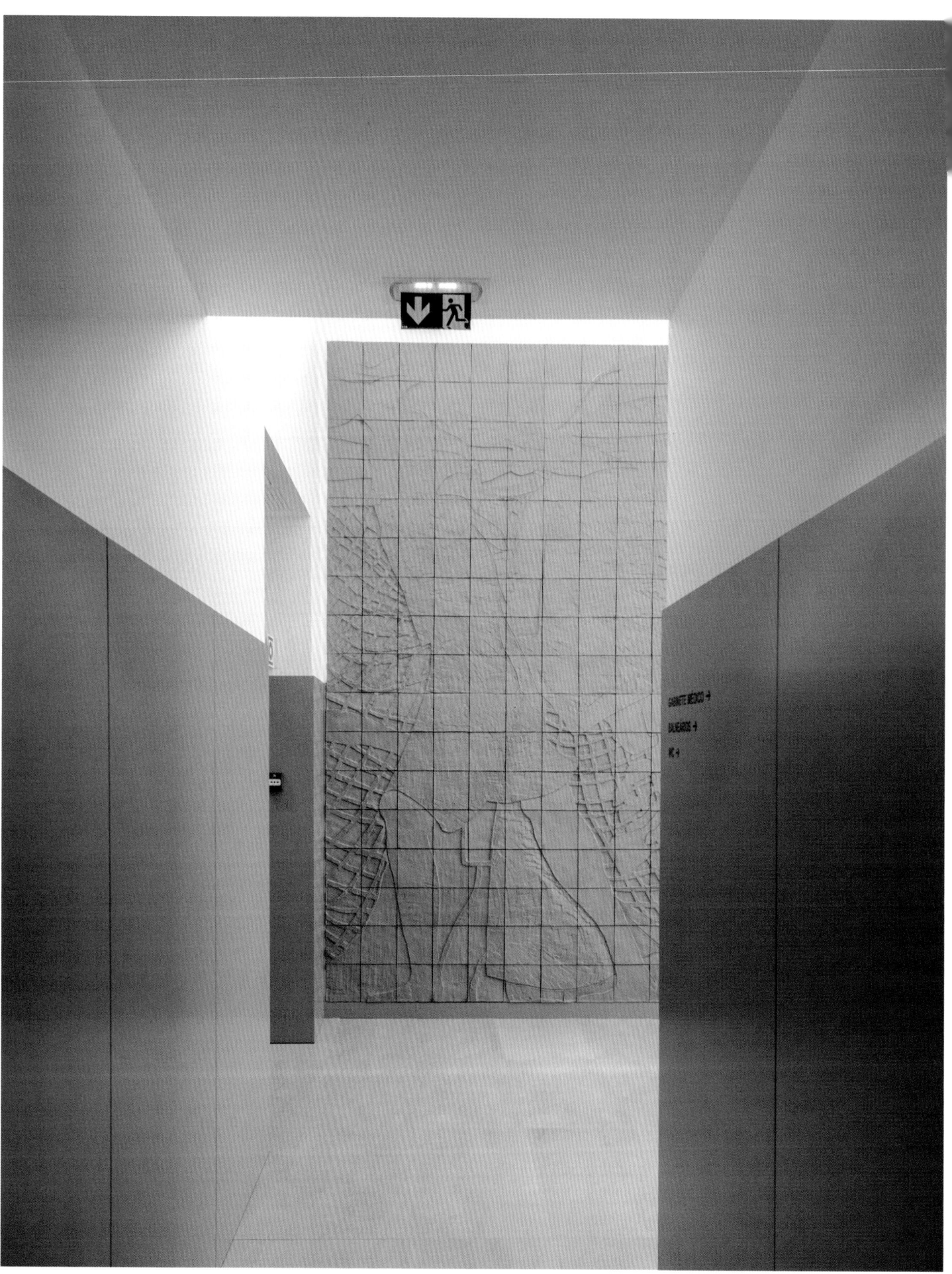

LOBBY SECTION

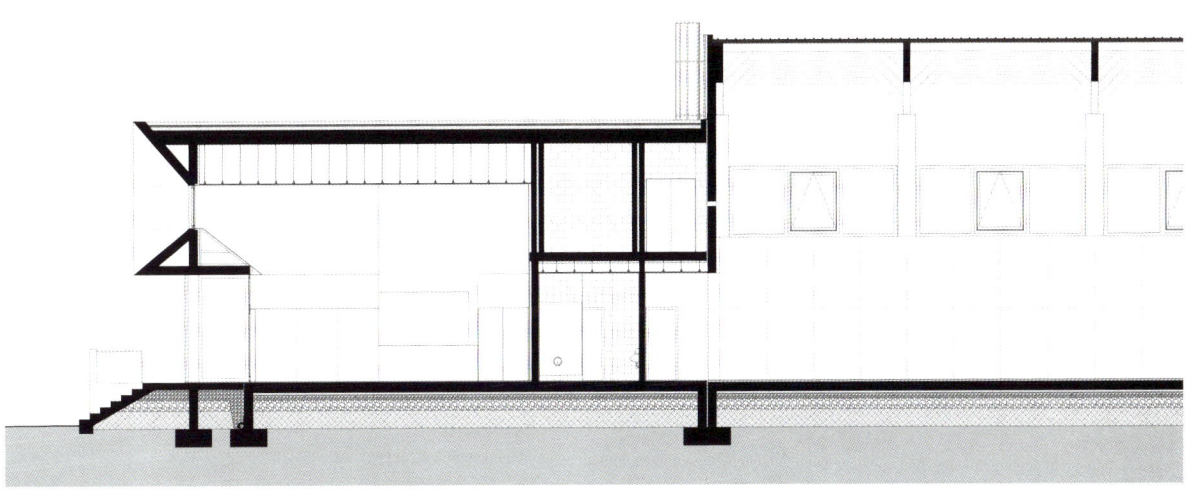

Echoing the Ocean

Curtain Wall

WENZHOU CENTURY PARK CULTURE CLUB
CULTURAL
STONE, STEEL, GLASS
WENZHOU, CHINA
2018

This project, located on a stretch of long and narrow land, is poised to conform to the dynamic trend of the green axis. Rocks have been placed along the street, forming an unusual interface and breaking the monotony of the streetscape.

The solid wall that runs through the west side of the whole building serves as the main shear supporting wall. In addition, three steel columns are concealed inside the stone curtain wall, and only one column in the room is exposed.

The secondary structure system adopts a combination of steel structure and curtain wall joist, and the façade glass curtain wall and roof aluminum plate share L-shaped curtain wall joist. The concept of the curtain wall design was introduced at the beginning of structural design, which not only saved structural costs but also simplified component size.

The internal space is largely open, with an average height of 33 feet (10 meters), and is simplified through the placement of equipment underneath the indoor floor.

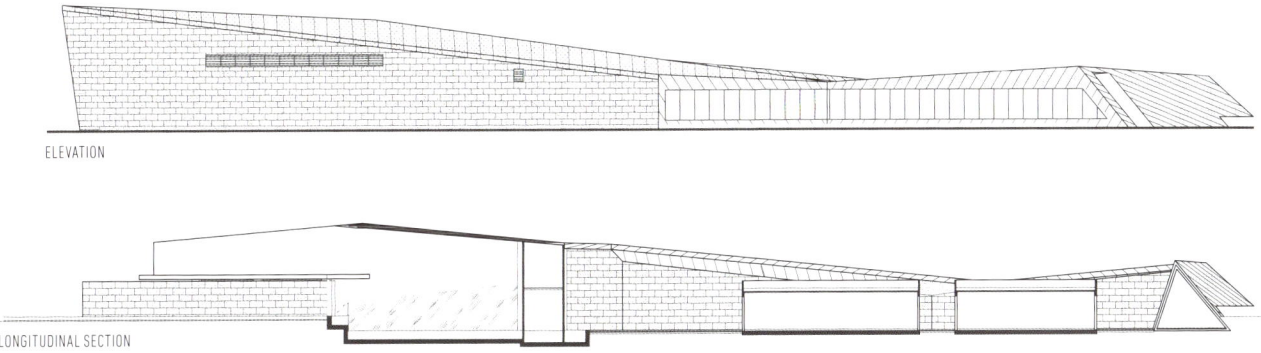

ELEVATION

LONGITUDINAL SECTION

Curtain Wall 91

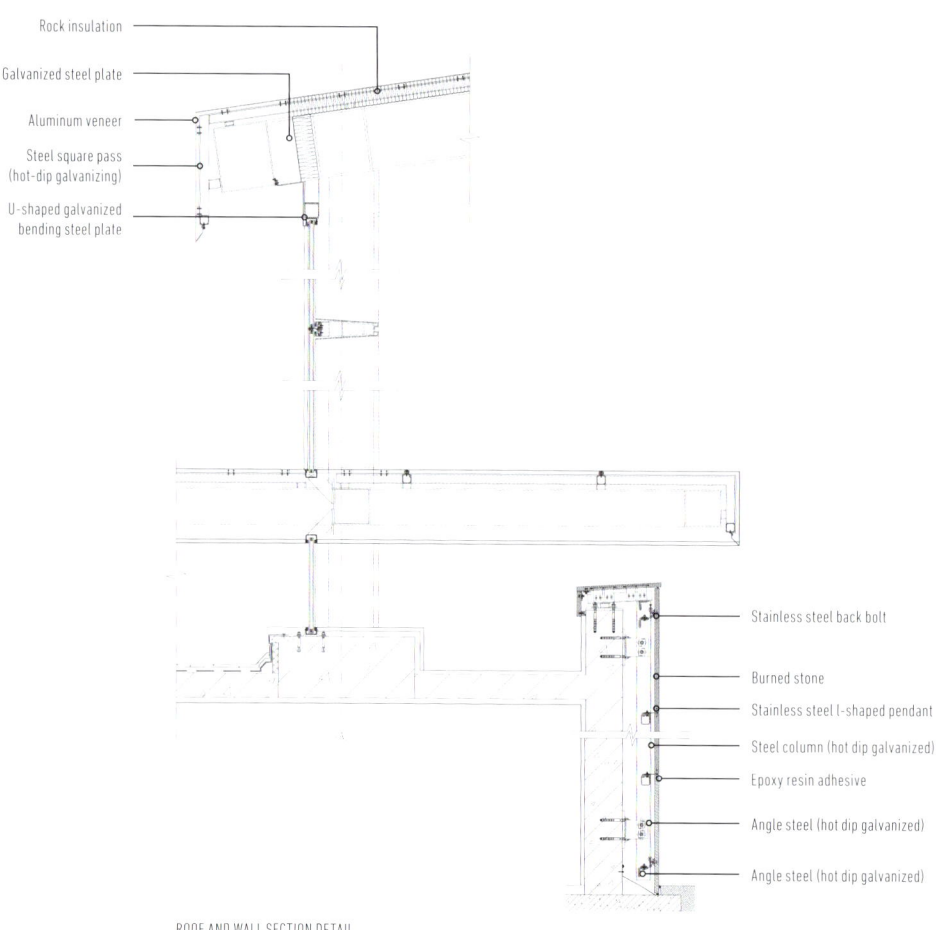

ROOF AND WALL SECTION DETAIL

Curtain Wall

Luxury in Stone

AKELARRE HOTEL
HOSPITALITY
STONE, WOOD
SAN SEBASTIAN, SPAIN
2017

Hotel Akelarre, located on a mountainside facing the sea, is the result of a more than forty years of identity and development. The original building, with its three-star Michelin restaurant, has been enhanced by the addition of five stone cubicles, designed by Madrid-firm, mecanismo.

The use of natural materials (stone, wood, and linen) adds a warmth to the overall design and gives the hotel an unequivocal personality. The stone gives an added texture at once both contemporary and ancient.

The cubicles emerge from the hillside, and contain the 22 hotel rooms, a snack bar, lounge, a wellness center, and a wine tasting room. The square dark-stone cubicles provide a contemporary touch, while the interior basks in the warmth of light wood tones, clear lines, and a muted palette.

The interior of the lounge boasts wooden cladding with an ocean-gray carpet and with sleek modern furniture, evoking a relaxed yet contemporary feel, while a bespoke stone fireplace provides a focal point, enhanced by a stone panel that runs the length of the room.

Stone feature walls are used to great effect throughout the hotel, especially when paired with a striking sculpture. The wellness space has been differentiated by a curved glass enclosure, and includes a luxury swimming pool, sauna, and Turkish bath. The dark tones of the stone cladding evoke a sense of luxury or of being inside your own private sea cave when contrasted with the blue of the water, and coupled with the view to the outside ocean.

Luxury in Stone 99

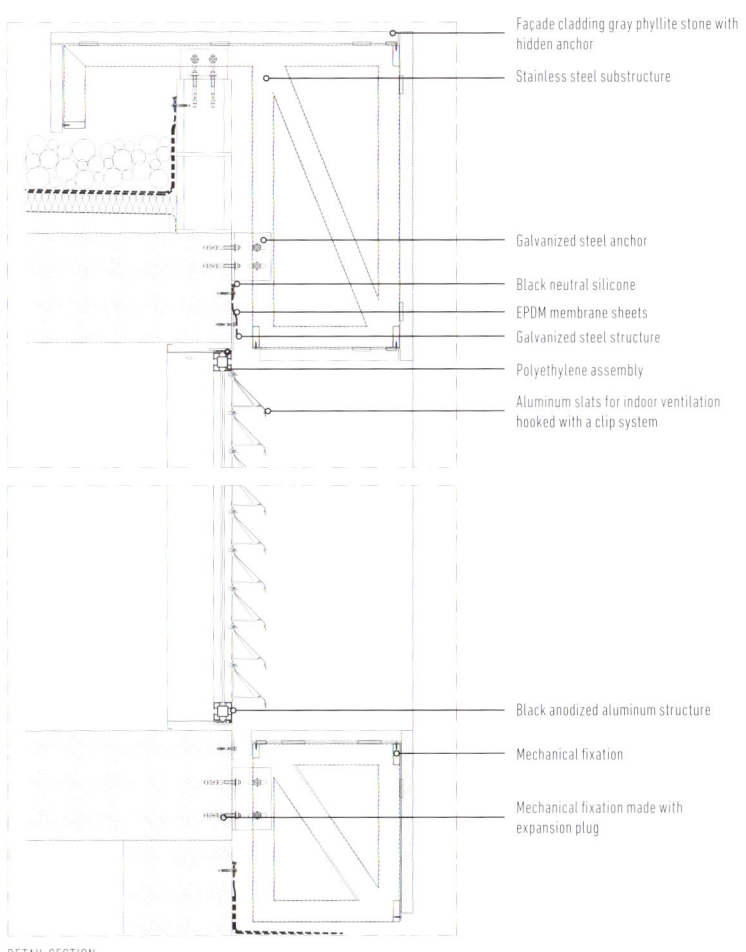

- Façade cladding gray phyllite stone with hidden anchor
- Stainless steel substructure
- Galvanized steel anchor
- Black neutral silicone
- EPDM membrane sheets
- Galvanized steel structure
- Polyethylene assembly
- Aluminum slats for indoor ventilation hooked with a clip system
- Black anodized aluminum structure
- Mechanical fixation
- Mechanical fixation made with expansion plug

DETAIL SECTION

Luxury in Stone   101

Elegant Lake Views

BÜRGENSTOCK HOTEL
HOSPITALITY
STONE, CONCRETE, GLASS
OBBÜRGEN, SWITZERLAND
2017

This imposing building forms the centerpiece of the Bürgenstock Resort Lake Lucerne, blending in naturally with the existing buildings. The overall hotel facility has been simplified from two buildings to one single, large, L-shaped building. This intervention created a spacious, public north terrace between the main building of the Bürgenstock Hotel and the Hotel Palace. The unimpeded access to the terrace via the Bürgenstock Piazza affords spectacular views over Lake Lucerne.

The new hotel sits prominently on the Bürgenberg rock, with its façade encased in pale gray-brown Sellenberg limestone, selected for its similarity to the natural Bürgenberg rock. From afar, the building looks like it has been hewn from the rock itself. The stone façade contrasts mainly with the metal elements such as the post-and-mullion glazing and the punch windows. Like the striking canopies, these are an elegant bronze color.

All 102 rooms face the lake and have a bay window as a central detail. The bathroom is connected to the sleeping area through a transparent glass sliding door. The bathtub offers—just like the bay window—a direct view over the lake. All rooms have their own gas fireplace, wardrobe, minibar, and desk. The materials were selected to harmonize with each other and draw the style of the entire hotel from the public areas into the private ones.

The project team succeeded in creating a fantastic combination of Swiss tradition and the modern, complete with responsible use of natural resources. The historical values have been maintained and the legendary history brought back to life.

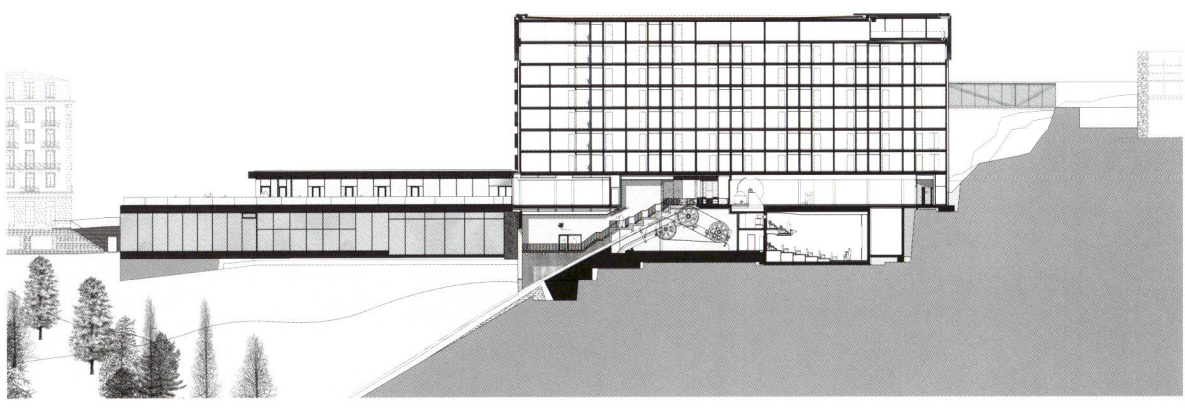

SECTION

Elegant Lake Views 107

Brick Stone Metal Wood

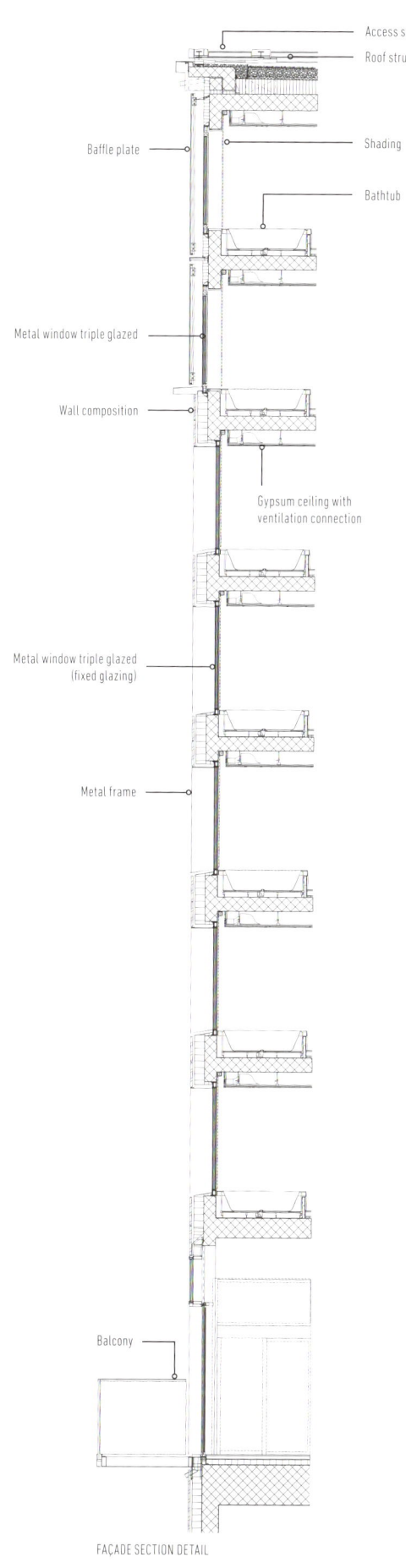

FAÇADE SECTION DETAIL

- Access system
- Roof structure
- Shading
- Bathtub
- Baffle plate
- Metal window triple glazed
- Wall composition
- Gypsum ceiling with ventilation connection
- Metal window triple glazed (fixed glazing)
- Metal frame
- Balcony

Elegant Lake Views   111

Pre-Hispanic Splendor

HUAYACÁN
HOSPITALITY
STONE, CONCRETE
JIUTEPEC MORELOS, MEXICO
2017

This 40-bedroom hotel in Jiutepec Morelos, Mexico is constructed from local stone, which contributes to its pre-Hispanic look. The result is a combination of old-world style and feel but with a contemporary modern flavor. The hotel comprises five volumes which are set on old poultry farm platforms, a nod to the region's rural past. Each volume is separated by courtyards, and set around the pool area, giving a feeling of space.

Access to the center of the hotel is via a narrow corridor under a concrete marquee, framed by three mature trees. Inside the marquee the stone is presented in different patterns, forming a pleasing whole. All the corridors are open to the elements, taking advantage of the clement weather conditions, and also allowing a greater sense of contact with the landscape. The corridor ceilings have an undulating white arc pattern.

Construction was largely comprised of load-bearing walls with a coverage of beams and arc-shaped concrete, resulting in a solid architecture with a light and neutral appearance. The interior design is simple, with white polished concrete floors, a neutral white palette and regional tile details.

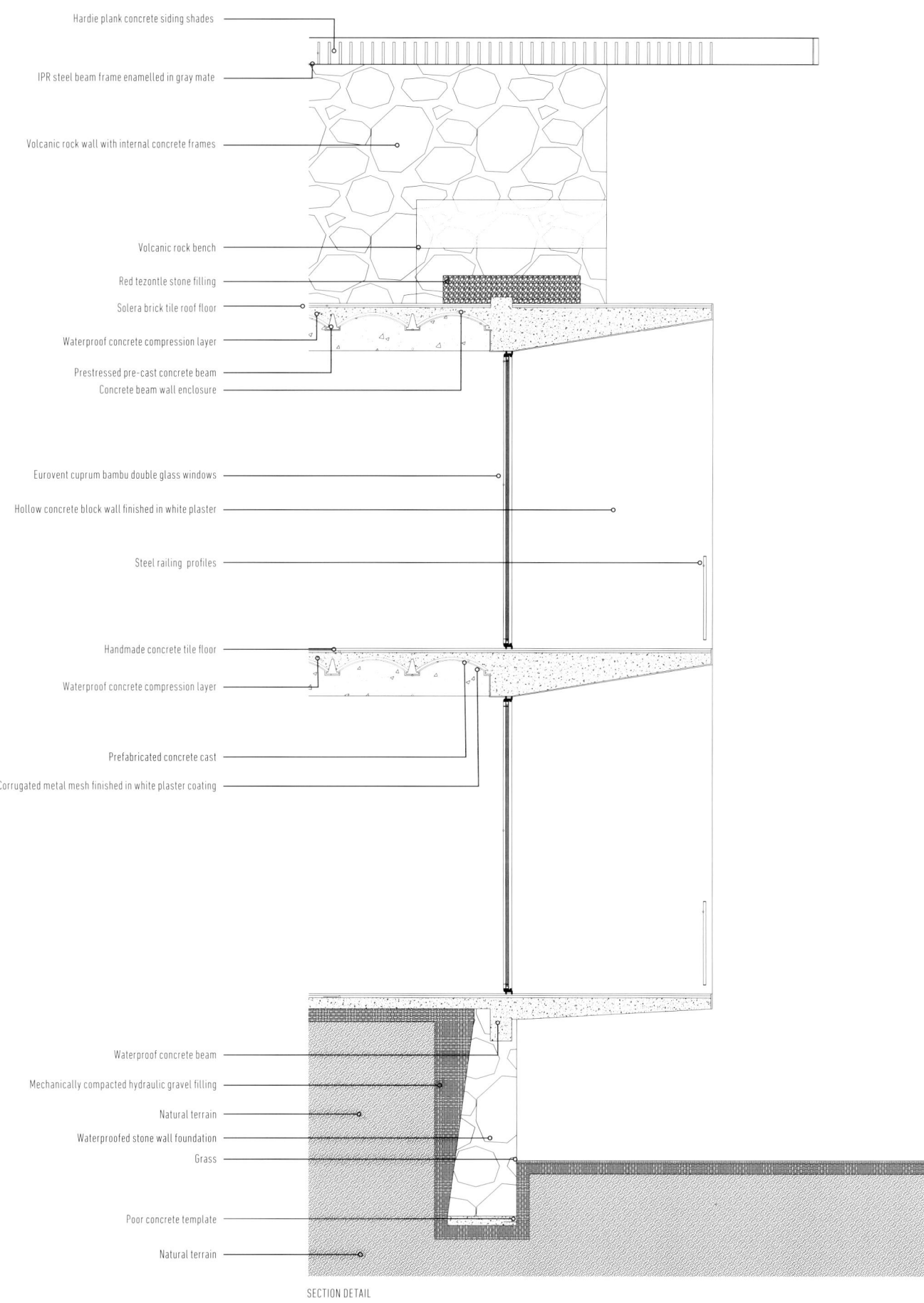

SECTION DETAIL

Pre-Hispanic Splendor

METAL in construction usually brings up images of steel, but there are other more traditional metals, such as copper or zinc, that are still used in modern buildings by contemporary architects alongside steel.

Looking at the Heroic Age

MUSEUM OF TROY
CULTURAL
CORTEN STEEL, CONCRETE, GLASS
ÇANAKKALE, TURKEY
2018

The archeological site of Troy is a UNESCO World Heritage Site for its outstanding universal value as a site that has witnessed various civilizations for more than 4,000 years, and has been a significant influence on the development of the European civilization, arts, and literature over two millennia.

The winning design for the museum was inspired by the idea of an "excavated artifact." The architects created a robust cubic form, and wrapped it in a metal (Corten) coated shell that will rust in time, evoking the connection between past and present.

The visitor descends into the cube along a wide ramp, leaving the Trojan landscape behind, and arrives at an underground band encircling the rust-red exhibition cube.

The design conceals all supportive functions underground on one floor. The large exhibition space is located within the cube and divided into four floors and a terrace accessible by ramps. The exhibition is divided into four floors: Troad and its Cultures, Bronze Age Troy, Iliad and the Troad in the Classical Age, History of Archeology at Troy.

The history of archeology at Troy has delivered a rich knowledge about the site and the Trojan landscape. The Homeric epic has an immense influence on cultural imagination. The exhibition brings these two forces together to create an understanding of the layers of settlement in Troy and its political and cultural impact in history. The terrace offers a preview of the site of Troy as well as other significant sites in the Troad as mentioned in *The Iliad* and excavated by archeologists for decades.

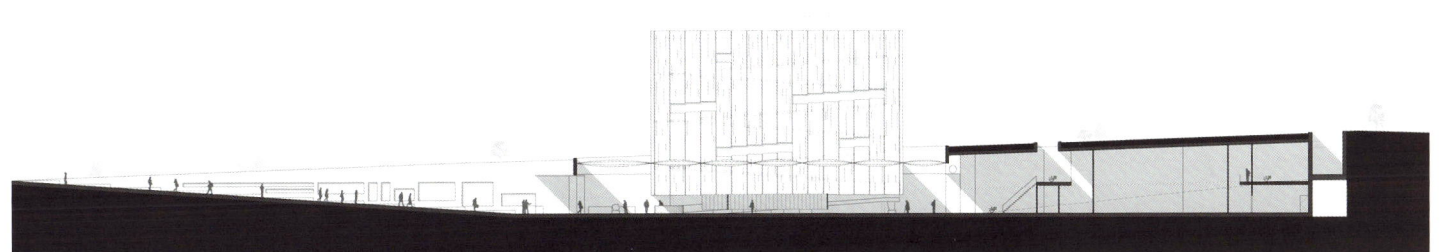
LONGITUDINAL SECTION

Looking at the Heroic Age 125

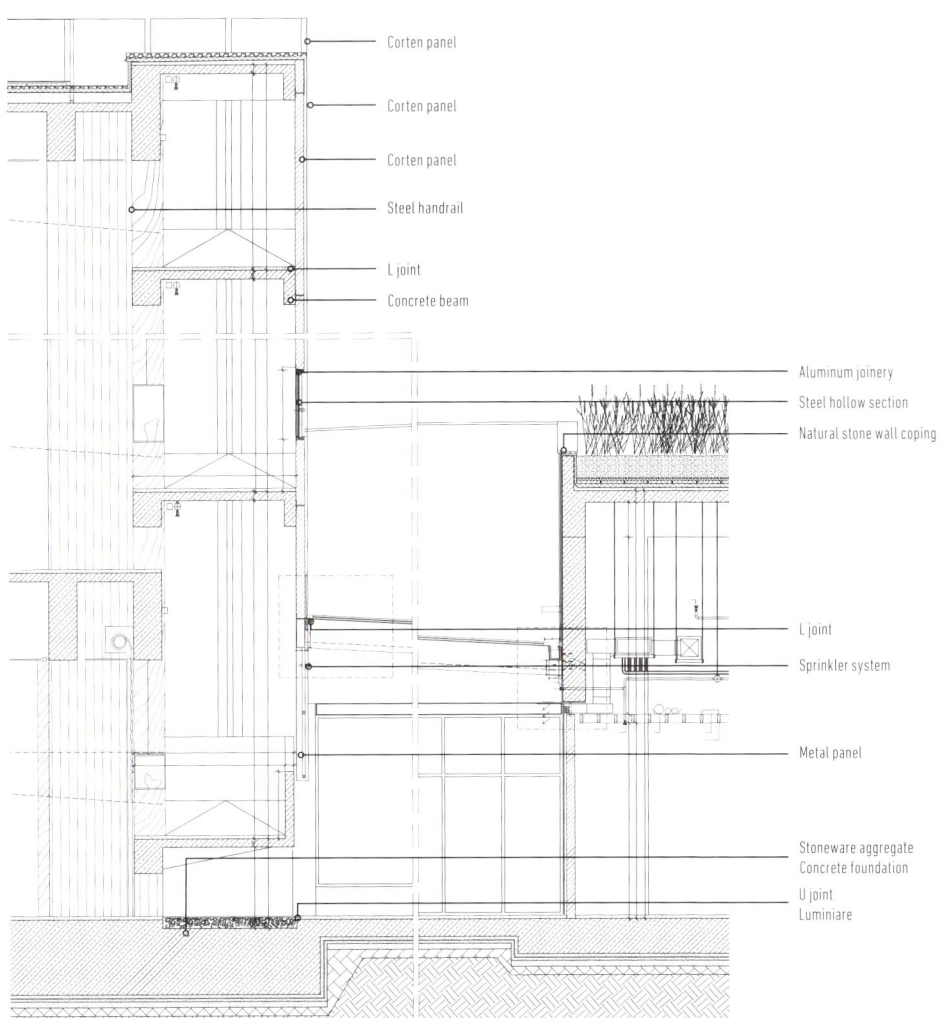

SECTION DETAIL

Looking at the Heroic Age

Blending In

NEW ENTRANCE HALL OF MUSEUM OF THE MIDDLE AGES
CULTURAL
ALUMINUM, ZINC, CONCRETE, WOOD
PARIS, FRANCE
2018

The old entrance to the Cluny Museum, the Museum of the Middle Ages and Roman thermal baths, had outgrown its ability to receive high numbers of visitors. In response, the new visitor center was designed to provide adequate services fitting to a national museum. The new addition restores the museum's presence on the street, and creates a fusion between its layers of history.

The building rests on a few micropiles, authorized by the archeology. By retaining the perception of the original structure, the new construction preserves the legibility and the succession of the old silhouettes. The cladding is made of cast aluminum modules of uneven dimensions and reliefs, contrasting with the stony masses of the remains.

The three façades are decorated with large flat areas of metal guipure, with a motif borrowed from an example of carved stone lace within the museum.

The connection with the old Roman annex, one of the oldest houses in Paris, is made by a set of footbridges that partially overlook and protect the remains of the Caldarium.

The interior organization consists of three platforms, expanding the museum's reception facilities. The internal staircase wraps around the large volume of the hall and inviting a perception of space, as well as indicating the path to the discovery of the museum's exquisite collections.

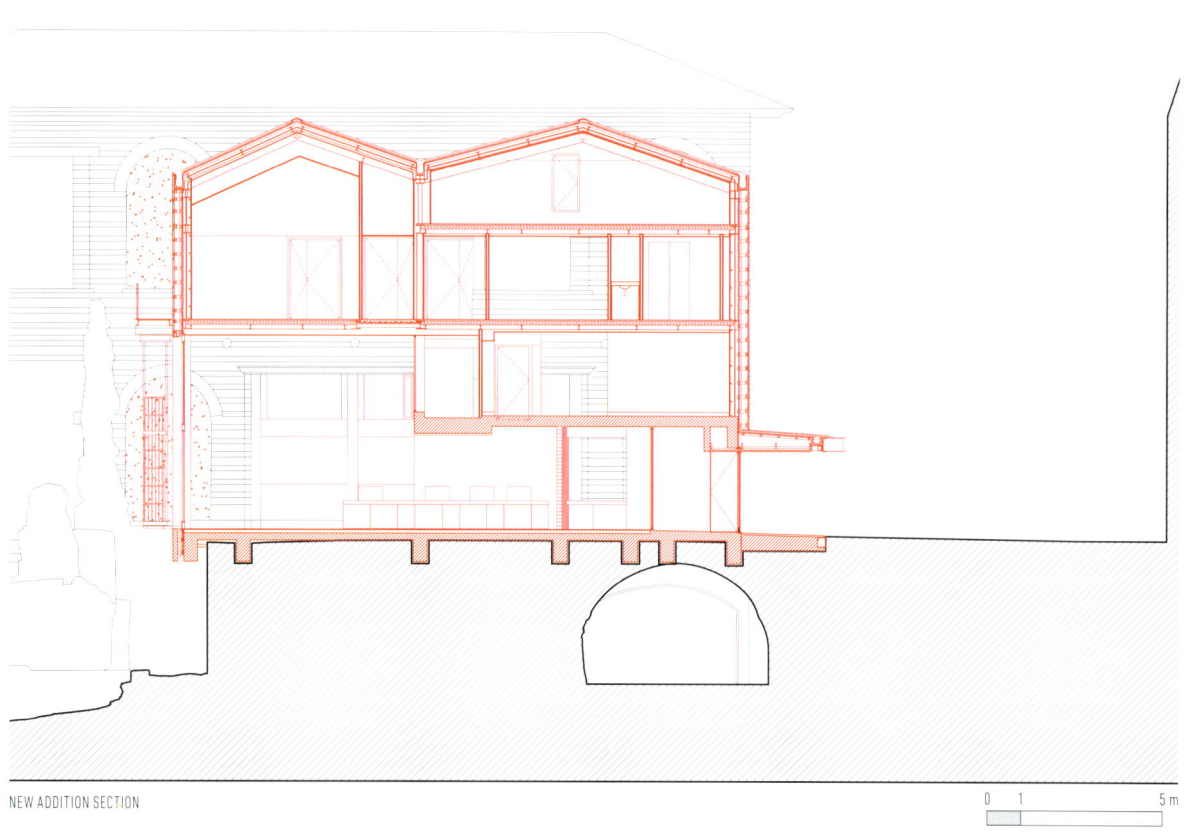

NEW ADDITION SECTION 0 1 5 m

136   Brick Stone **Metal** Wood

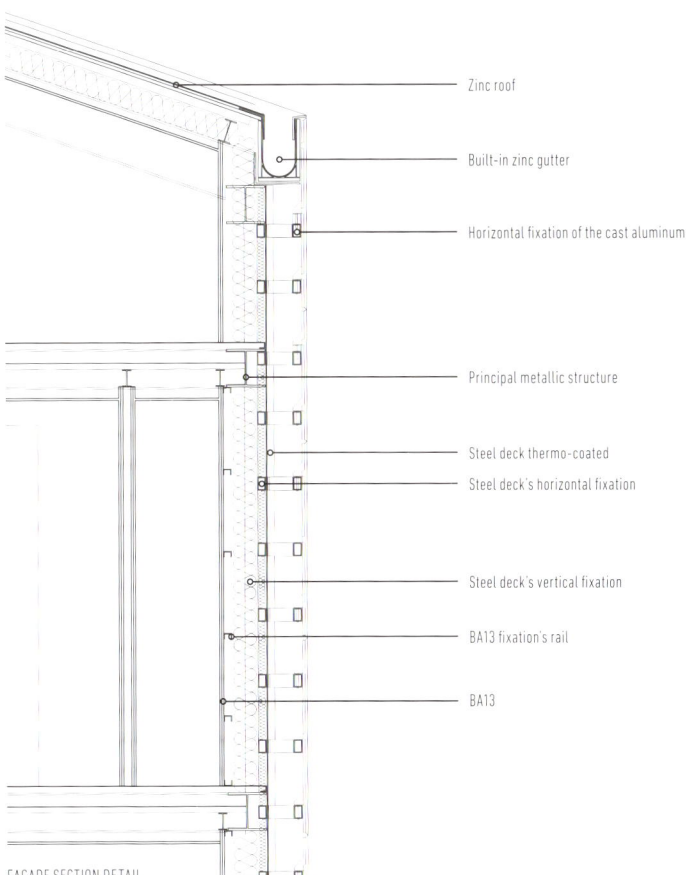

- Zinc roof
- Built-in zinc gutter
- Horizontal fixation of the cast aluminum
- Principal metallic structure
- Steel deck thermo-coated
- Steel deck's horizontal fixation
- Steel deck's vertical fixation
- BA13 fixation's rail
- BA13

FAÇADE SECTION DETAIL

Blending In 137

Revisiting Ancient Tradition

SHUI CULTURAL CENTER
CULTURAL
COPPER, CONCRETE, WOOD
GUIZHOU PROVINCE, CHINA
2017

With its iconic and distinct shape, the Shui Cultural Center stands as a new contemporary landmark which pays homage to the local Shui culture and traditional architecture.

The iconic shape of the cultural center was inspired by the Shui language, which has its own unique system of pictographs, following the shape of the character for "mountain." The façade pattern is also inspired by Shui's traditional characters, starting again from the basic triangular shape of the mountain, which is repeated to evoke the character for "rain."

The site is surrounded by water on three sides, appropriate as "Shui" means water. The use of copper by the Shui inspired the architects to use perforated copper steel plates to cover the building. The pattern makes the plates lighter—a thin skin which creates a contrast with the heavy concrete structure—breaking the sunlight to create a dramatic effect once inside.

The concrete is marked strongly by a wooden pattern, given by the pine quarterdecks. Pine wood is one of the most common materials in the local area and the contemporary concrete structure echoes the local traditional wooden architecture.

The building itself comprises three main parts, which combine all the functions of the tourist-cultural center. The first is the ritual hall, which with its sharp edges, strong colors and narrow space aims to create a strong first impression on visitors. The second section welcomes visitors with less dramatic tones and serves as a reception hall. The third section has a more conventional space and includes all the main utilitarian functions, from a visitor and service center to a cafeteria.

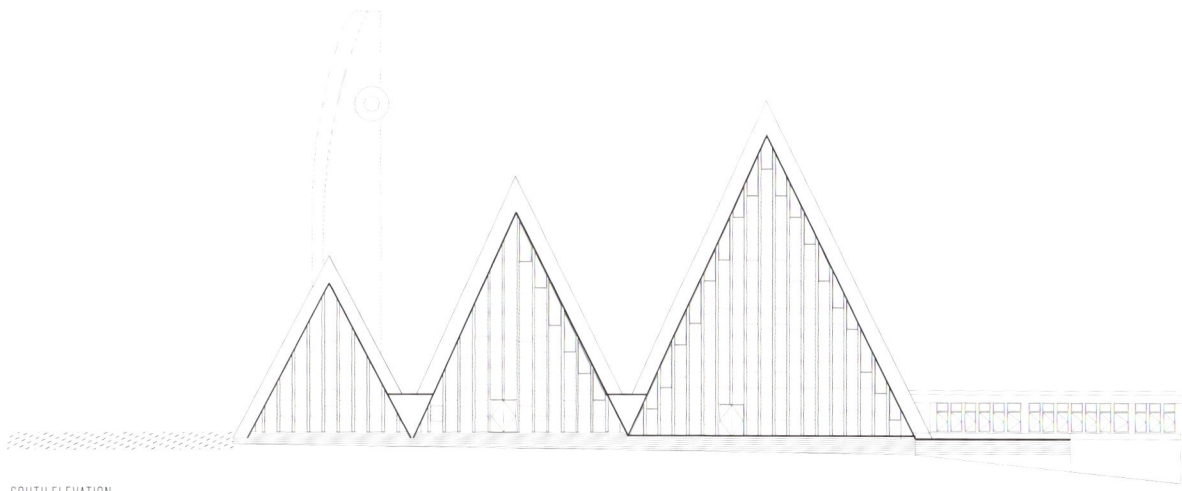

SOUTH ELEVATION

Revisiting Ancient Tradition 143

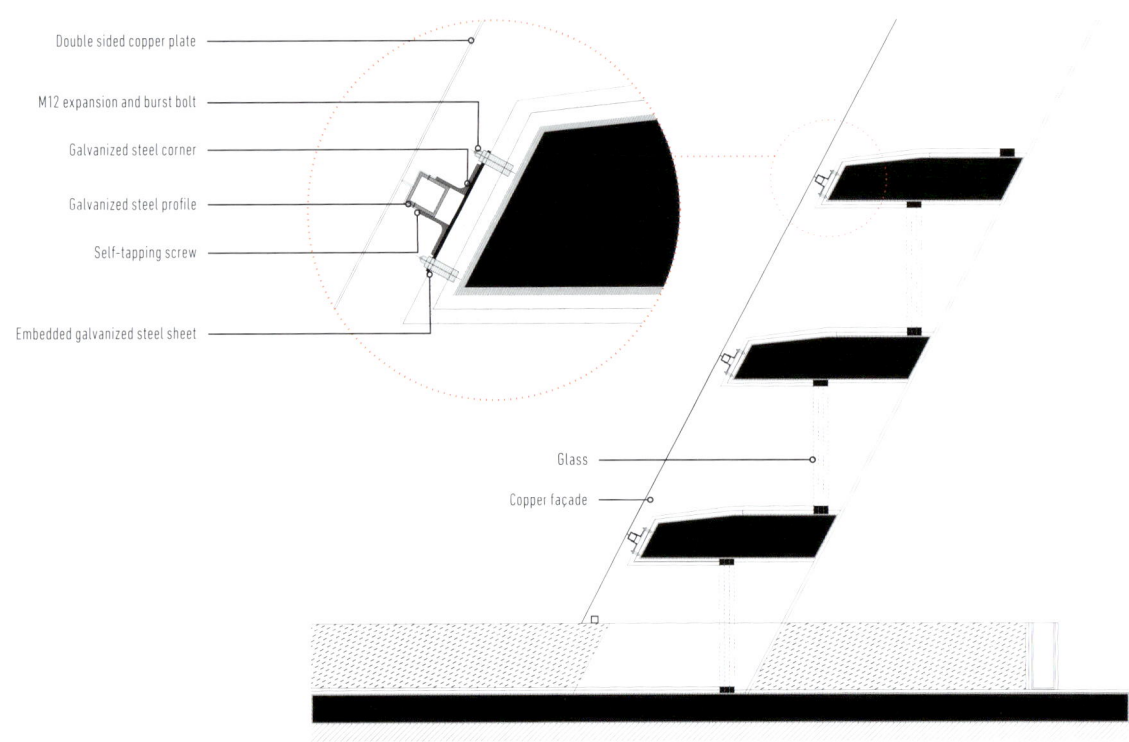

FAÇADE DETAIL

- Double sided copper plate
- M12 expansion and burst bolt
- Galvanized steel corner
- Galvanized steel profile
- Self-tapping screw
- Embedded galvanized steel sheet
- Glass
- Copper façade

144 Brick Stone **Metal** Wood

A-Framing the View

SIERRA
HOSPITALITY
CORTEN STEEL, WOOD, BRICK
LONE TREE, UNITED STATES
2017

Sited on the crest of a hill overlooking Lone Tree, Colorado, Sierra is a restaurant housed in a purpose-built building designed to capture views and create a dynamic dining experience. Combining classic building forms like the A-frame and the courtyard, it makes a timeless, modern statement.

Fourteen steel A-frames create the signature form of the restaurant. The downhill leg of the A-frames extend outside the building envelope to a concrete buttress wall on the hillside below. Half of the A-frames enclose the main dining room, while the second half continue over an exterior deck. In this area, the corrugated, Corten roofing becomes perforated to provide dappled light during the day, and acts as a lantern at night.

On the opposite side of the building, a large courtyard creates a sheltered outdoor dining experience. Generous roof overhangs provide shelter along the building, and steel window frames on the courtyard wall provide a wind block while maintaining views.

Salvaged red brick continues from the exterior to the interior, while wood floors, hammered copper tables, and exposed Corten steel complete the warm materials palette. The restaurant's interior is anchored by the open kitchen. The wood-fired pizza oven and grill provide focal points, while two layers of copper counters allow guests to see the chefs at work.

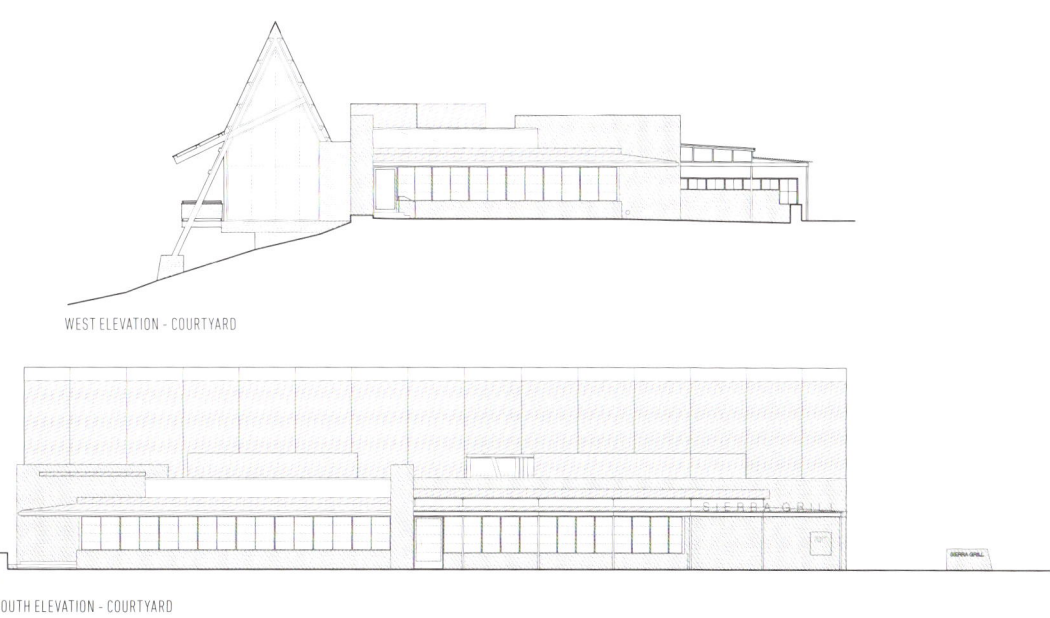

WEST ELEVATION - COURTYARD

SOUTH ELEVATION - COURTYARD

A-Framing the View

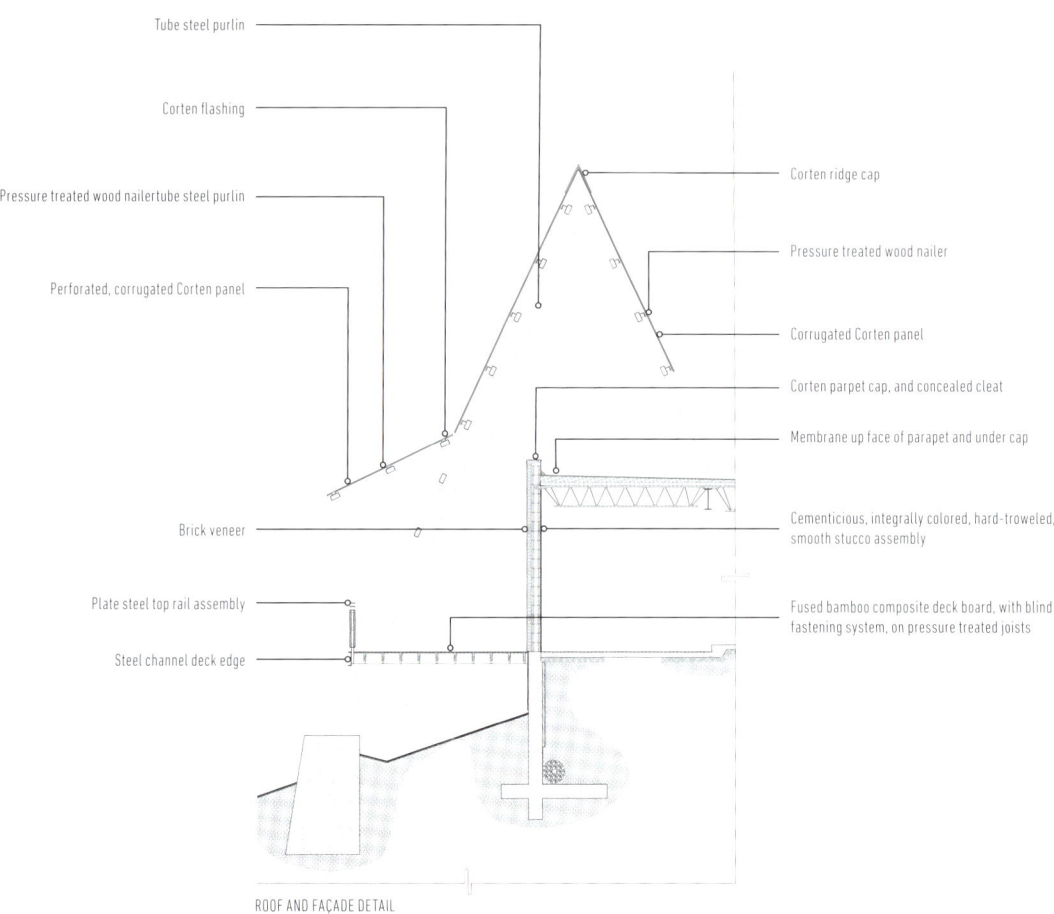

ROOF AND FAÇADE DETAIL

WOOD as a natural product has been present in construction since the earliest shelters built by man. Bestowing an elemental warmth, wood is able to transform its surrounding environment. With its associated flexibility, it continues to be used in diverse ways by modern architects.

Private Access

HOUSE BEHIND THE ROOF
RESIDENTIAL
WOOD, CONCRETE, TILES
KRAKOW, POLAND
2018

Situated in a densely developed housing estate, the house is hidden from the access road and its northern neighbors behind its striking roof surface. Hence its name: House Behind the Roof.

The green roof has a 45-degree slope, with the northern part covered with succulents while photovoltaic cells are placed on the sunny southern part. The house is constructed with laminated timber with diagonally guided elements and undercuts at the edges of the roof. At the connection of the two roof slopes windows let in a significant portion of natural light.

The architects paid special attention to how the building would change over time, and what the effect would be on the materials used. The elevation of the building was covered with Western Red Cedar planks. This wood in time it will take on a noble patina and change its color to silver-gray. The green roof with its hardy succulents does not require special care.

The house has been designed on a rectangular plan with two floors. The garage is placed inside the house, on the ground floor, near to living room and kitchen, allowing convenient access in winter months. Also located on this level are a guest room, toilet and utility rooms. On the first floor there is master bedroom with dressing room and bathroom, two bedrooms for children, children's bathroom, and kid's play room.

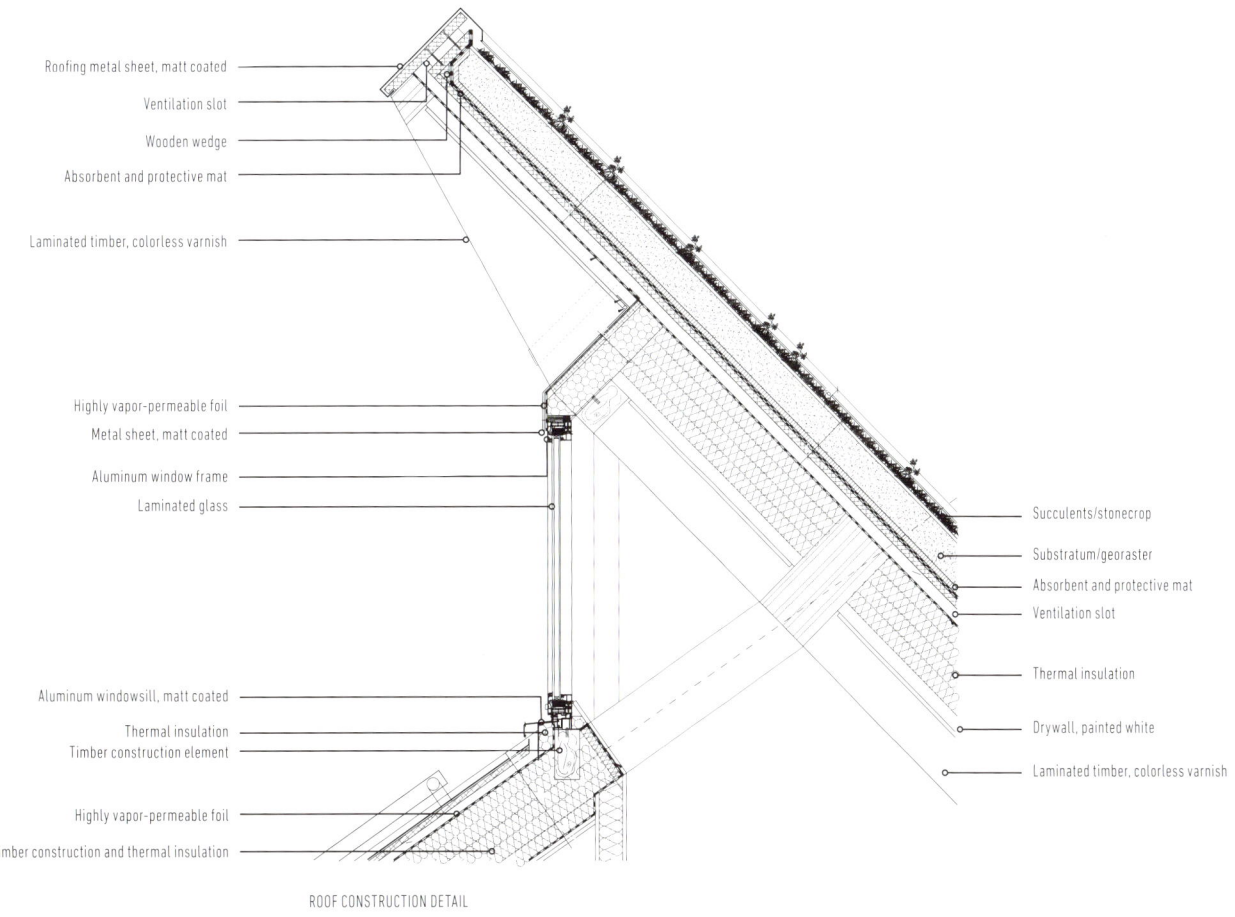

ROOF CONSTRUCTION DETAIL

Private Access

New Façades

IBIZA GRAN HOTEL
HOSPITALITY
WOOD, STONE, GLASS
IBIZA, SPAIN
2018

The ambitious renovation project of the existing building aimed to add 30 new rooms, and completely remodel the hotel's façades.

The façades are particularly important in the unrelenting Ibiza sun, as they provide protection against the strong sunlight. They also have the added bonus of providing a unified image for the hotel. The long vertical timber slats provide an updated and contemporary look to the building, and combine well with the light stone used. Seen from an angle, the façades form a stepped series of vertical planes, where the displacement creates small balconies for the hotel rooms.

The internal design relies on the warmth of the light-colored wood from timber-clad walls to the wooden flooring, which combines with the stark white walls to provide a soothing and cooling effect.

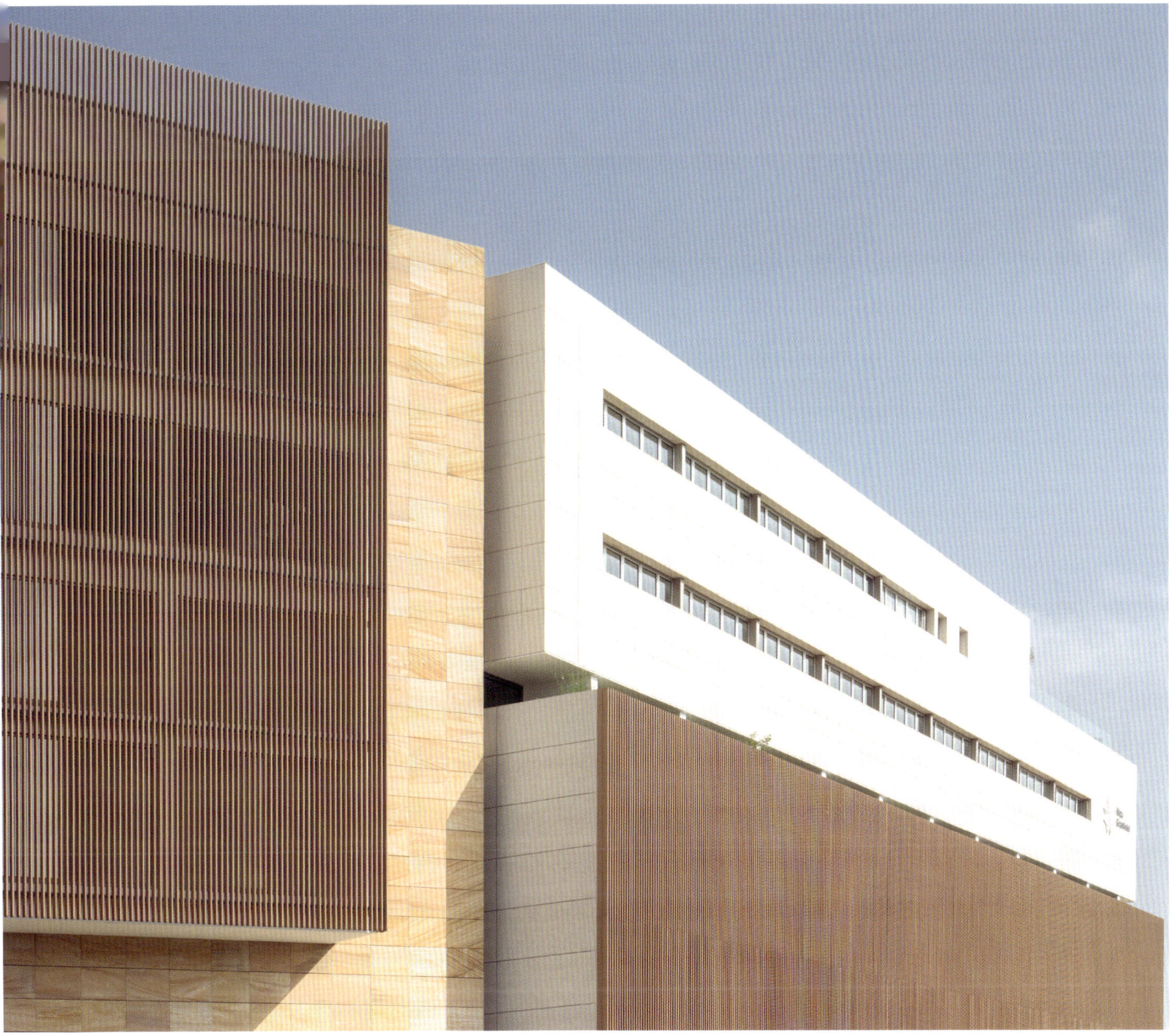

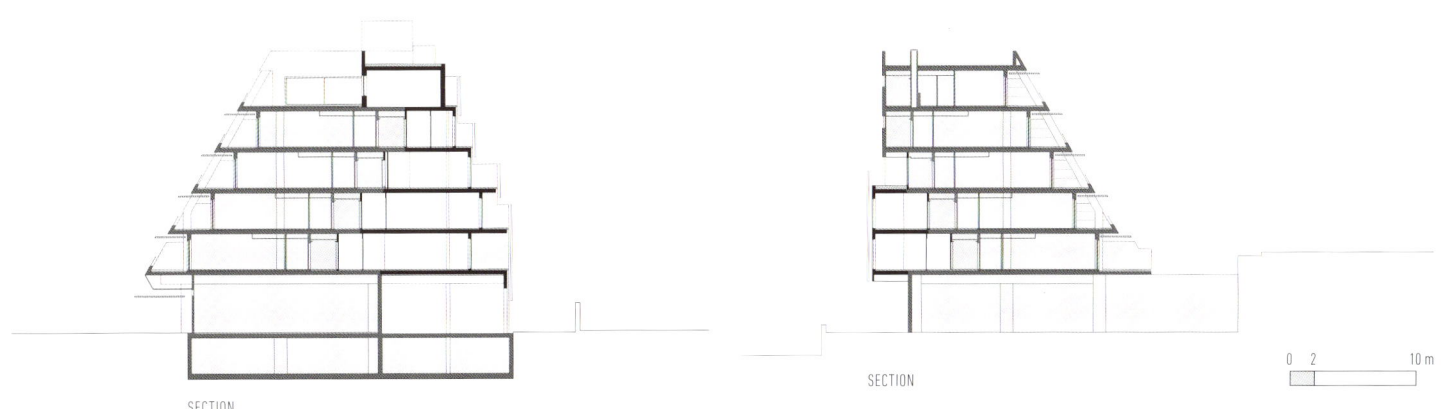

SECTION

SECTION

0 2   10 m

New Façades   165

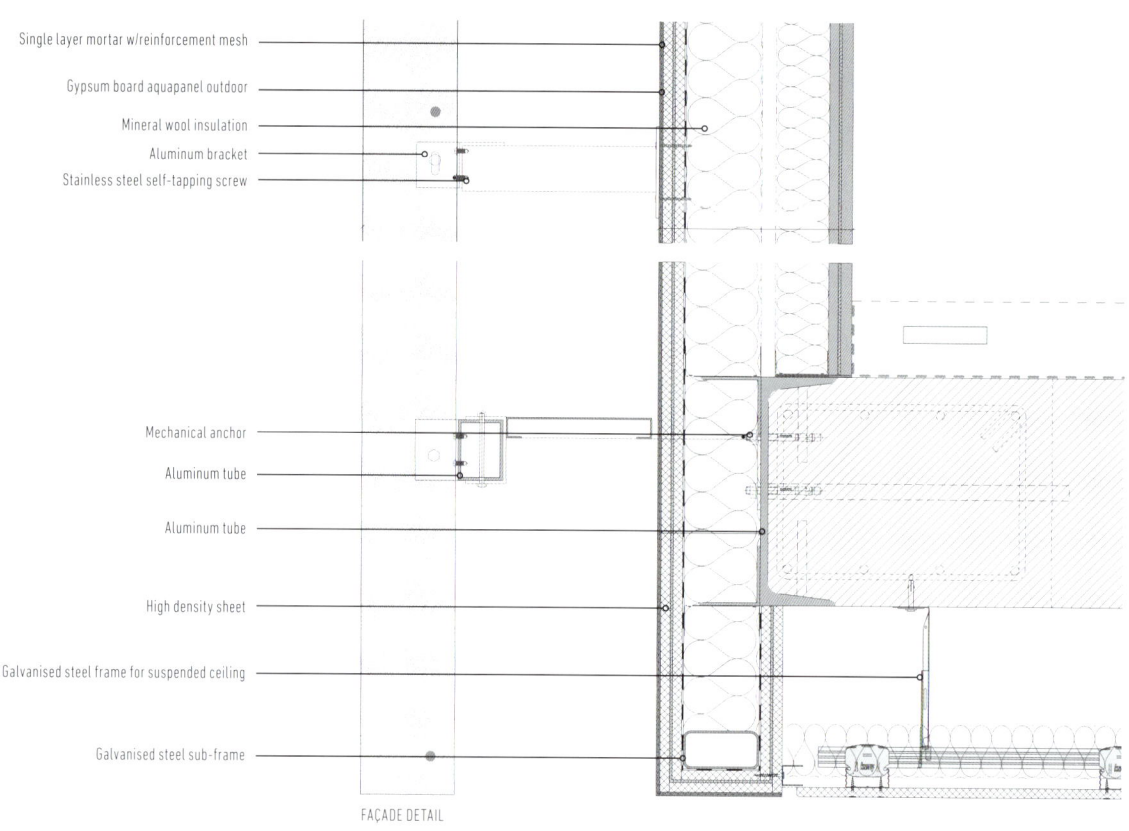

- Single layer mortar w/reinforcement mesh
- Gypsum board aquapanel outdoor
- Mineral wool insulation
- Aluminum bracket
- Stainless steel self-tapping screw
- Mechanical anchor
- Aluminum tube
- Aluminum tube
- High density sheet
- Galvanised steel frame for suspended ceiling
- Galvanised steel sub-frame

FAÇADE DETAIL

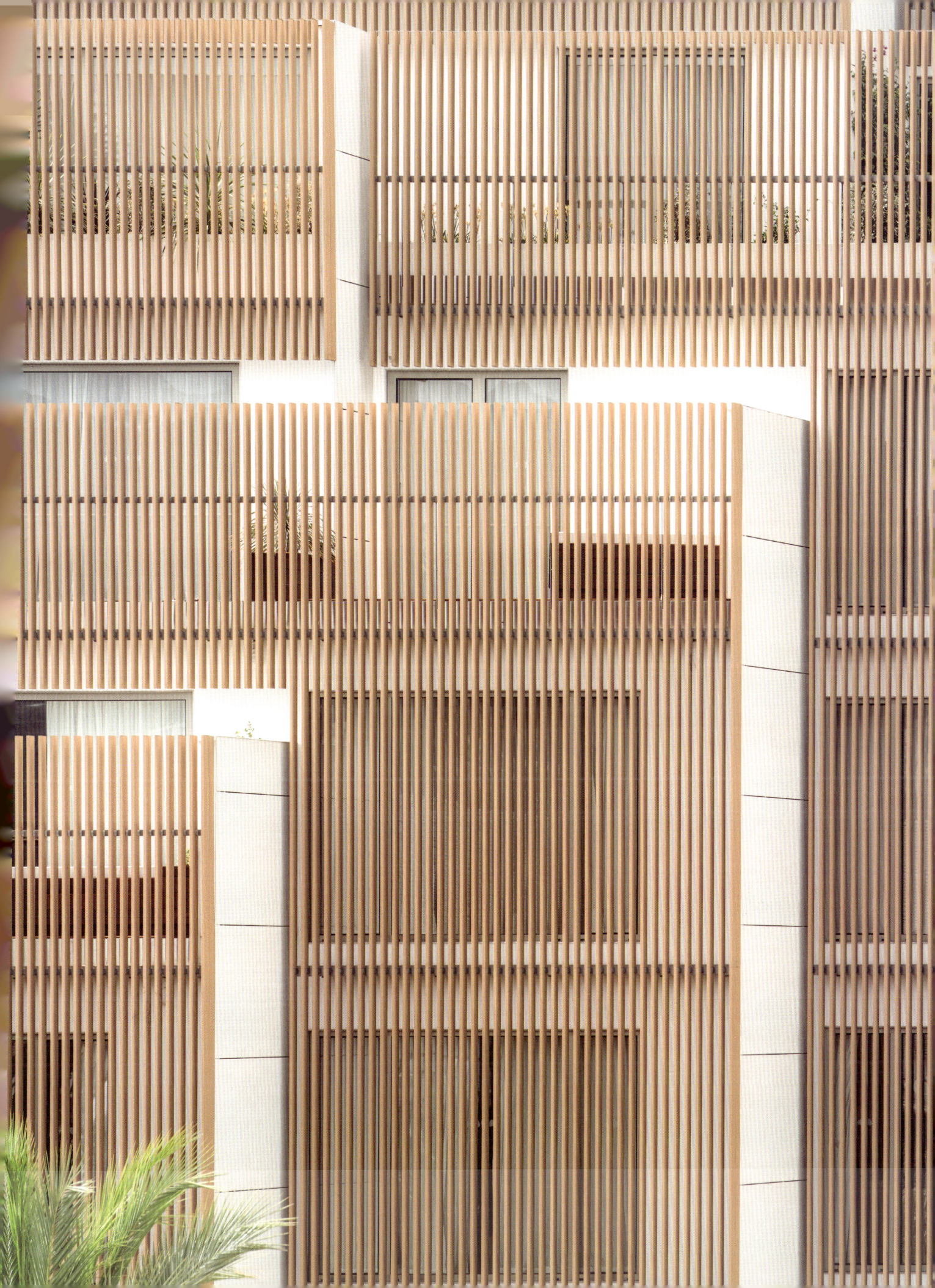

The Heart in Ikast

HJERTET
INSTITUTIONAL
WOOD, CONCRETE, GLASS
VESTERGADE, DENMARK
2018

Hjertet (the Heart), as the project is called, provides an expansion to the International School Ikast-Brande, also designed by the same architect in a similar style. The new building rises like a dark wooden monolith above the green hillocks, yet the interior is open and filled with natural light from various windows, and bathed in warm wood tones, with dark gray floors.

The addition features a central square with a performance stage, which distributes users out to the various rooms in the building. One wing holds the school's teaching rooms, which in the afternoons and evenings can be changed to multi-rooms and art workshops for associations and evening schools. For young sports enthusiasts, the street sports hall is particularly interesting, as it is designed so as to retain a sense of being outdoors. There is also a café with a service kitchen, and a shop area where organic groceries from a local socio-economic initiative as well as handcrafts by blind producers are sold.

On the first floor of the building there are various large and small rooms for activities or cultural events. The surrounding activities landscape is designed around sustainable drainage principles and includes facilities for different active pursuits, such as parkour or skateboarding.

The Heart is a very open and accessible building, and reinforces its capacity to act as a social gathering point through its use of many small, informal venues and seating areas.

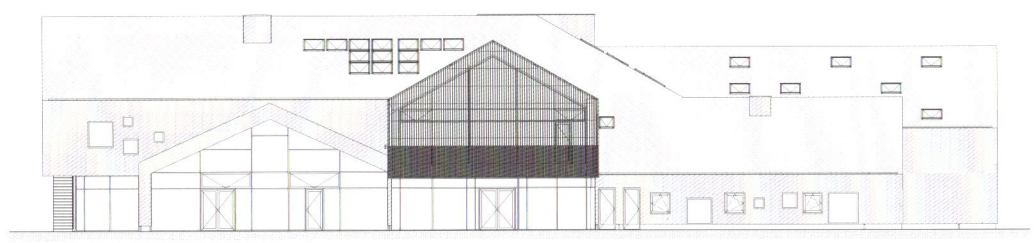

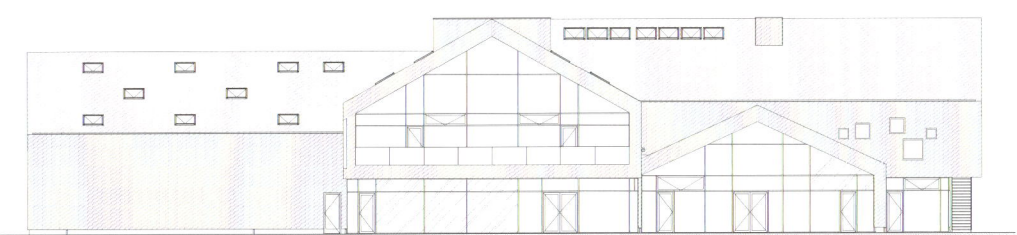

EAST ELEVATION

WEST ELEVATION

The Heart in Ikast

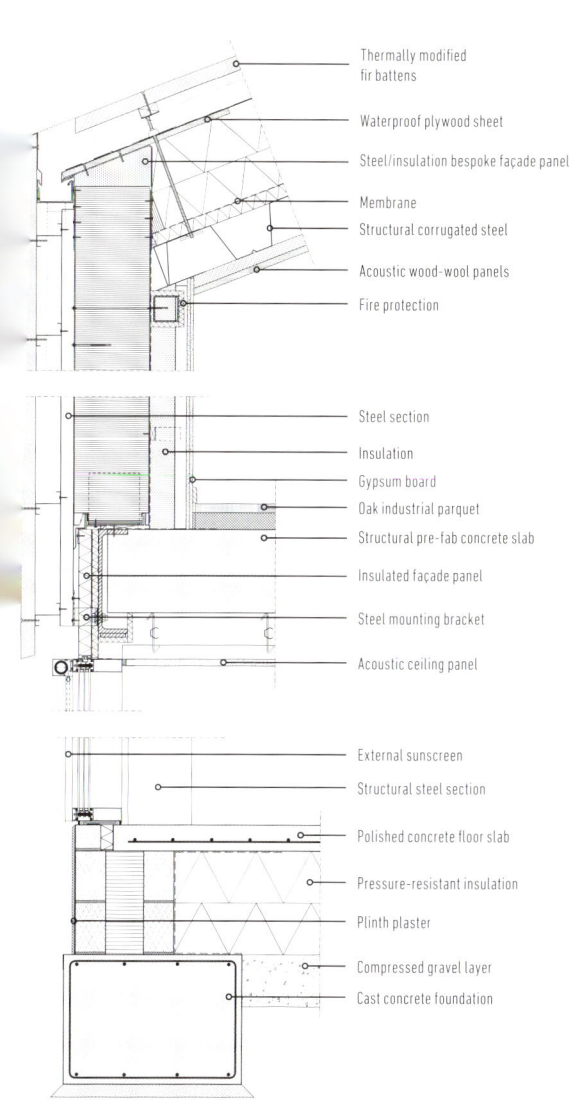

FAÇADE DETAIL

- Thermally modified fir battens
- Waterproof plywood sheet
- Steel/insulation bespoke façade panel
- Membrane
- Structural corrugated steel
- Acoustic wood-wool panels
- Fire protection

- Steel section
- Insulation
- Gypsum board
- Oak industrial parquet
- Structural pre-fab concrete slab
- Insulated façade panel
- Steel mounting bracket
- Acoustic ceiling panel

- External sunscreen
- Structural steel section
- Polished concrete floor slab
- Pressure-resistant insulation
- Plinth plaster
- Compressed gravel layer
- Cast concrete foundation

The Heart in Ikast   179

Stacking the Terraces

INTER CROP OFFICE
COMMERCIAL
WOOD, STEEL, ALUMINUM
BANGKOK, THAILAND
2018

In rethinking a multistory office building for a leading agricultural trading company, the designers envisioned a new workplace experience, using the concept of a "rice terrace" that in turn becomes an architectural embodiment of the company's philosophy and business operation.

A series of overlapping cantilevers and terraces was created by shifting the mass of each floor slightly differently to each other. The simplicity of the stacked terrace was a sophisticated solution and allowed the local flora and fauna of the outdoor landscape to permeate into the interior.

These "rice terraces" serve as a flexible common area for the office workers. The overhanging eaves provide ample of shading for gathering, sharing, or working outdoors during the day.

The aluminum vertical fins of the façade are designed to provide dynamic shading from the sun's harsh rays. These vertical slats help to significantly reduce the reliance on air-conditioners by filtering over-exposed light and excess heat from the sun.

The sheer mass of the "rice terrace" acts as a natural insulation, helping to soften the direct heat gain. This results in the creation of a microclimate for the building which helps to neutralize the imbalanced climate conditions that may arise.

This new working environment is distinct from a typical office building, and not only creates a new working experience but also transforms the building into a green space in its surrounding context.

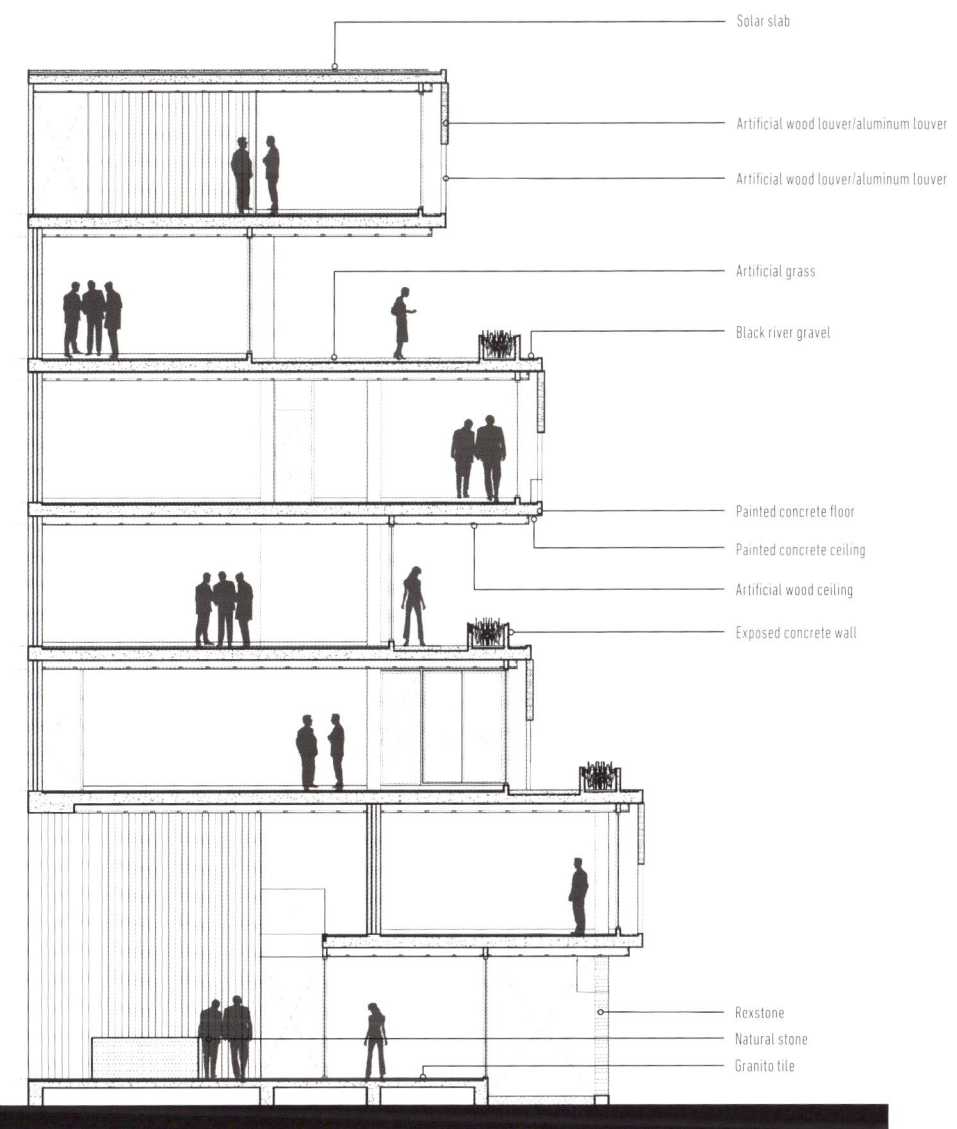

- Solar slab
- Artificial wood louver/aluminum louver
- Artificial wood louver/aluminum louver
- Artificial grass
- Black river gravel
- Painted concrete floor
- Painted concrete ceiling
- Artificial wood ceiling
- Exposed concrete wall
- Rexstone
- Natural stone
- Granito tile

SECTION

Stacking the Terraces   187

Rise Like a Phoenix

POLYVALENT HALL
INSTITUTIONAL
WOOD, STEEL, GLASS
LE VAUD, SWITZERLAND
2018

Le Vaud's new community hall with its unique pointed roof is designed to serve and bring together the entire village community. Just before its official opening, the building was entirely destroyed by fire, before being carefully reconstructed.

The main materials used in its construction are local timber, the standard metal sheeting typically used by farms in the area, and the sandblasted concrete of the existing school—simple materials that create a dialogue with the local surroundings.

The building uses the site's natural slope as the basis for its functional program. The multipurpose space of the hall itself is integrated with the natural terrain, while its indented roof outline echoes the roof shapes of the existing school buildings.

The geometry of the building's interior space does not exactly match the geometry of its outer envelope. These offset geometries are most clearly apparent in the triangular openings that structure the design of the building's gables.

Glazed triangular windows are set within larger-scale wooden triangles, protected by claustras which filter the light. The north and south façades, by contrast, are fully glazed and transparent, allowing natural light to flood in. The north façade, with the entrance, is very low and protected by a large overhang—shielding the reception forecourt from the typical weather conditions at the foot of the Jura Mountains. A similar overhang on the south side ensures effective solar protection.

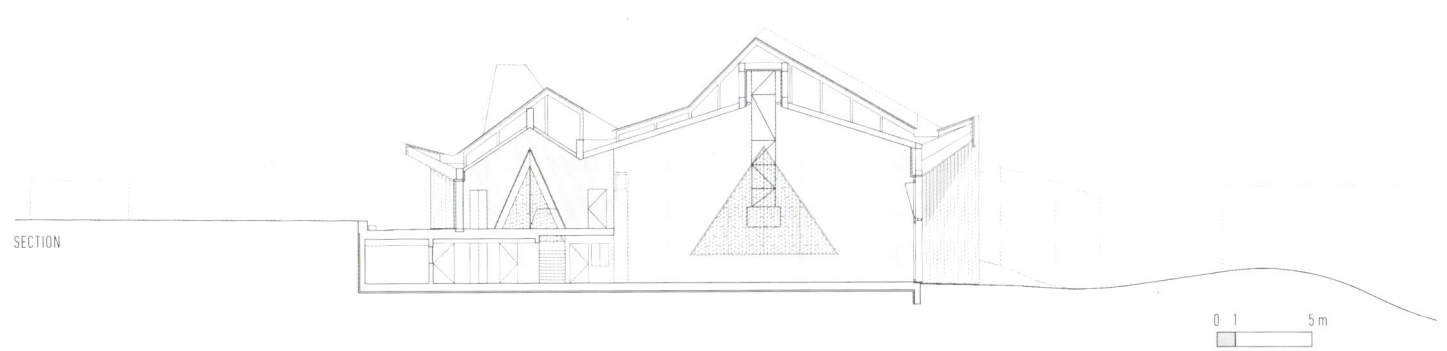

SECTION

0 1   5 m

Rise Like a Phoenix

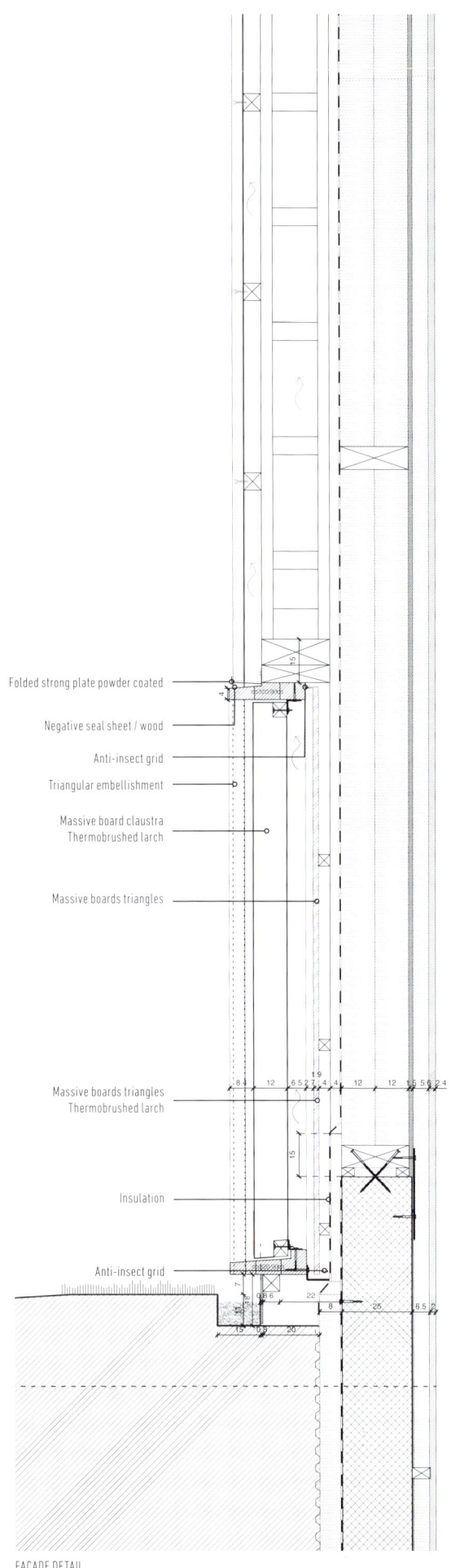

- Folded strong plate powder coated
- Negative seal sheet / wood
- Anti-insect grid
- Triangular embellishment
- Massive board claustra Thermobrushed larch
- Massive boards triangles
- Massive boards triangles Thermobrushed larch
- Insulation
- Anti-insect grid

FAÇADE DETAIL

Treasure Box

STEINHARDT MUSEUM OF NATURAL HISTORY
CULTURAL
WOOD, STONE, GLAZED TILES
TEL AVIV, ISRAEL
2018

This eye-catching modern design is the first museum of its kind in Tel Aviv, the Steinhardt Museum of Natural History. Not only does it house the vast natural history collections of Tel Aviv University and serve as a center for academic research for staff, but visitors are encouraged to visit the exhibitions, and are able to see the researchers at their work.

The impressively large building is environmentally friendly and housed within a striking architectural structure composed of a wooden-panel shell. The wooden "treasure box" is thermally insulated to afford complete climate control of the interiors. The interior is punctuated with openings to the outside, allowing natural light to penetrate the interior.

Floating above ground, the museum's entrance plaza and gathering lawn allow a seamless view of the gardens from the street level. The façade facing the street is an intriguing combination of stone, wood, and glass, while the western façade, resting on two large metal poles, presents a more brutalist concrete face.

The Steinhardt Museum of Natural History combines both exhibition spaces and research activities within a modern edifice wrapped with an insulated wooden shell. Above the main interior exhibition space on the building's upper levels lies the research laboratories for Tel Aviv University's staff.

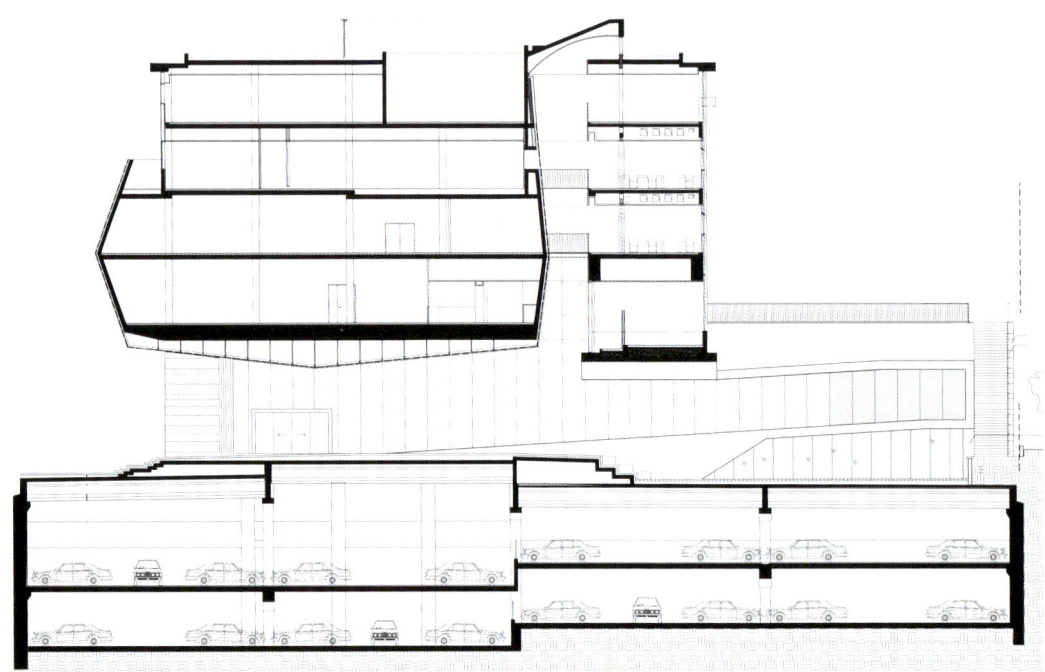

EAST-WEST SECTION

200 Brick Stone Metal **Wood**

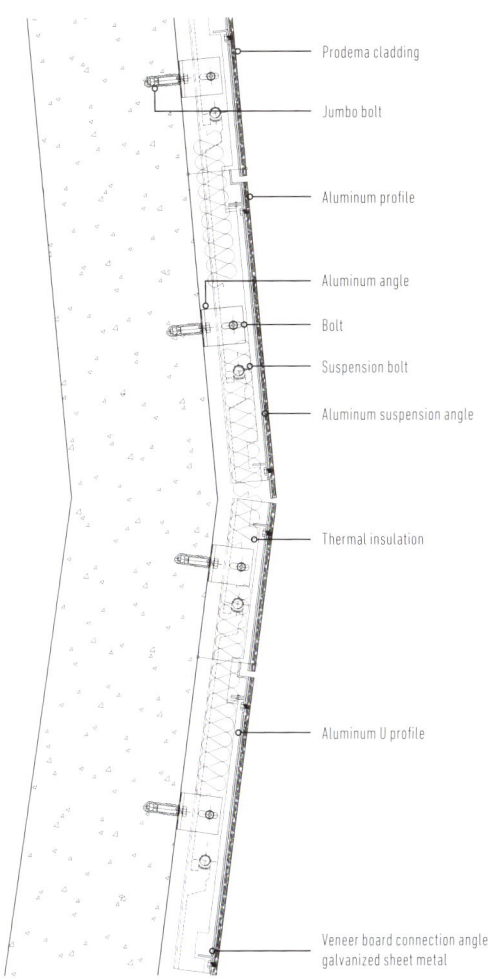

- Prodema cladding
- Jumbo bolt
- Aluminum profile
- Aluminum angle
- Bolt
- Suspension bolt
- Aluminum suspension angle
- Thermal insulation
- Aluminum U profile
- Veneer board connection angle galvanized sheet metal

FAÇADE DETAIL

Treasure Box   **201**

Student Rhythm

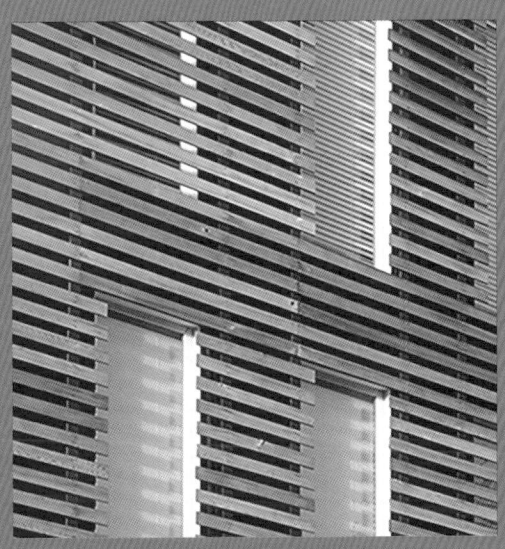

ARKANSAS BEAR CLAW
RESIDENTIAL
WOOD, CONCRETE, BRICK
FAYETTEVILLE, ARKANSAS, UNITED STATES
2017

This student-focused multifamily project occupies a trapezoidal site adjacent to the University of Arkansas in downtown Fayetteville.

The courtyards between building wings are extraverted engageable spaces capable of playing with topography, pedestrians, and drivers alike.

A layered palette of brick, naturally weathering cedar screens and siding, fiber cement board, and steel composes massive areas of dense five-story apartment construction containing 628 bedrooms and 228 units. The wings of the building are focused around preserved specimen trees and programed community amenity spaces. The tenant clubhouse is counterintuitively located mid-block, a glass box that provides a transparent threshold between the street and the pool courtyard. The origami roof form captures the entry and provides a roof deck that gives tenants the outdoor opportunity to straddle the public and private realm.

The Arkansas Bear Claw presents a uniquely scaled street experience that is at once an inviting urban rhythm and articulated building form along the Fayetteville hillside.

ELEVATION

Student Rhythm

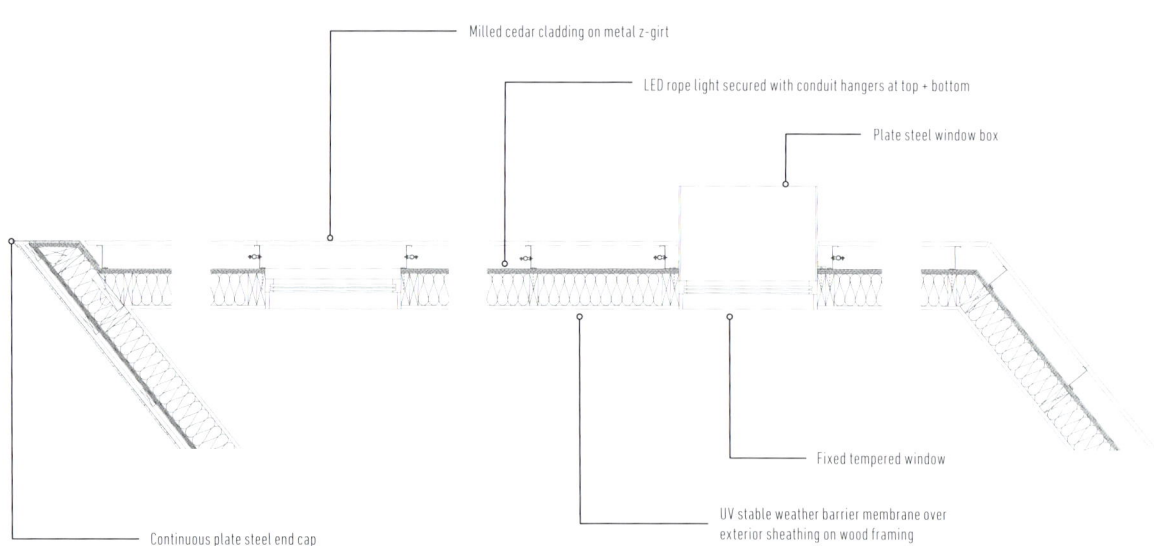

FAÇADE DETAIL

- Milled cedar cladding on metal z-girt
- LED rope light secured with conduit hangers at top + bottom
- Plate steel window box
- Fixed tempered window
- UV stable weather barrier membrane over exterior sheathing on wood framing
- Continuous plate steel end cap

Student Rhythm 207

Disappearing Trick

HOLMEN AQUATICS CENTER
CULTURAL
WOOD, STEEL, GLASS
ASKER, NORWAY
2017

Holmen Aquatics Center is designed as a continuation of Holmen beach, by protecting and reinforcing the site's natural qualities. The structure is designed to literally blend in with its surroundings, by way of a green roof, beneath which the building disappears into the slope. The grassy roof also offers spectacular views over nearby isles and reefs. Its front façade provides a vision of a wooden ship atop the blue watery waves.

The main entrance, reception, changing rooms, and swimming pool hall are located on the building's main level, rising over the lawn that slopes up to the building. The lower level houses the gym, multifunction hall, technical spaces, and staff facilities.

By employing Passive House methodology a longterm vision is established for energy use. The methodology encompasses construction methods, insulation and building form. Furthermore, energy reuse is stressed, especially with regards to water heating.

The building uses solar panels in addition to 15 deep geothermal wells that provide heat from the bedrock below the site and are also used for transferring excess heat down to the ground during summer months.

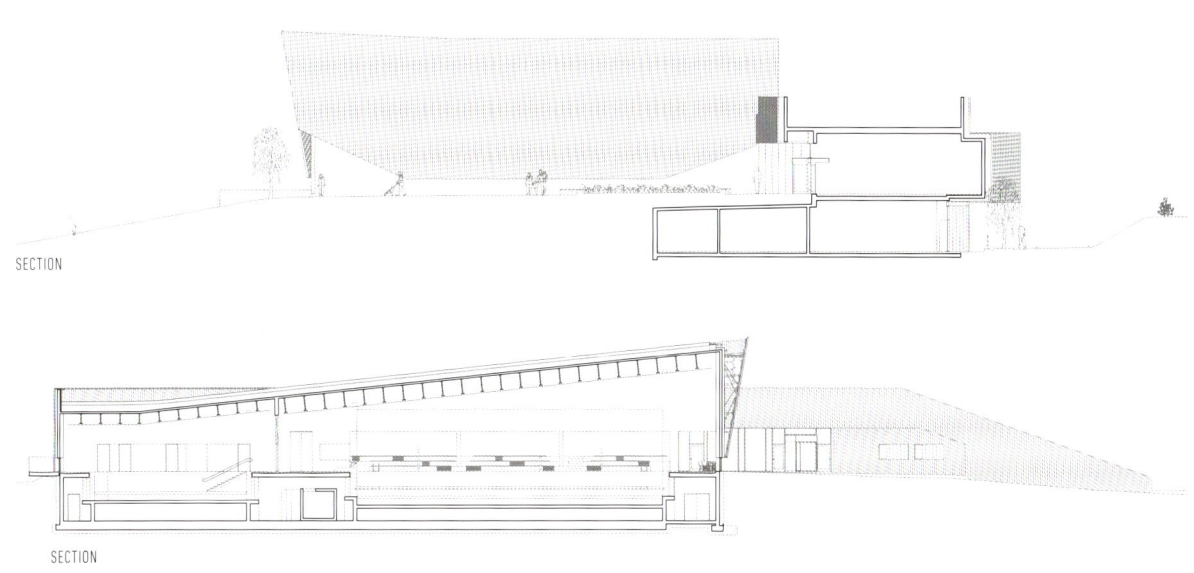

SECTION

SECTION

Disappearing Trick

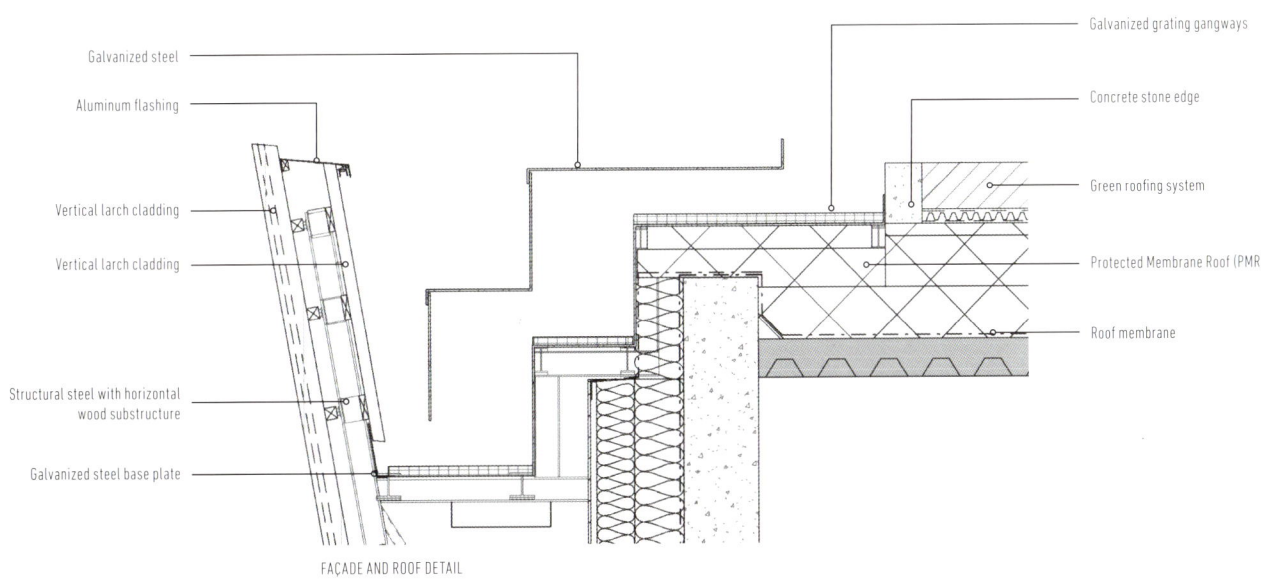

FAÇADE AND ROOF DETAIL

- Galvanized steel
- Aluminum flashing
- Vertical larch cladding
- Vertical larch cladding
- Structural steel with horizontal wood substructure
- Galvanized steel base plate
- Galvanized grating gangways
- Concrete stone edge
- Green roofing system
- Protected Membrane Roof (PMR)
- Roof membrane

Disappearing Trick

Planned Spontaneity

IMPLUVIUM COMMUNITY CENTER
CULTURAL
WOOD, CONCRETE, GLASS
CANTABRIA, SPAIN
2017

IMPLUVIUM may be summed up as comprising a large roof astride a laminated timber structure, replacing a former market demolished after a fire. The roof, together with four boxes that arise from the ground, is the necessary infrastructure that allows many events to happen, both spontaneous and programmed under the same ceiling.

The square plan is organized by the strategic location of the boxes and a courtyard that polarized the interior space. The courtyard is decentered to generate different scale spaces in the ground level. Also, its floor is lowered and moved in relation with the glass enclosure, generating outside shaded sitting areas for the summer time and an inside adjacent bench that receives the direct sunlight through the windows during the winter.

The ground floor is in continuity with the site level on the south, extending the public space toward the inside and visually connecting the whole building with the street. In addition, the strategy used allows the introduction of a mezzanine that hangs from the main structure and offers private spaces. Its location in the perimeter creates a lineal layout for the plan with a more individual scale, taking advantage of the natural light, filtered by the wooden lattice that surrounds the building.

The variety of particular situations in and by the courtyard, around the space, or in front of the building, generates a great flexibility of use without losing the specificity of the different scenarios.

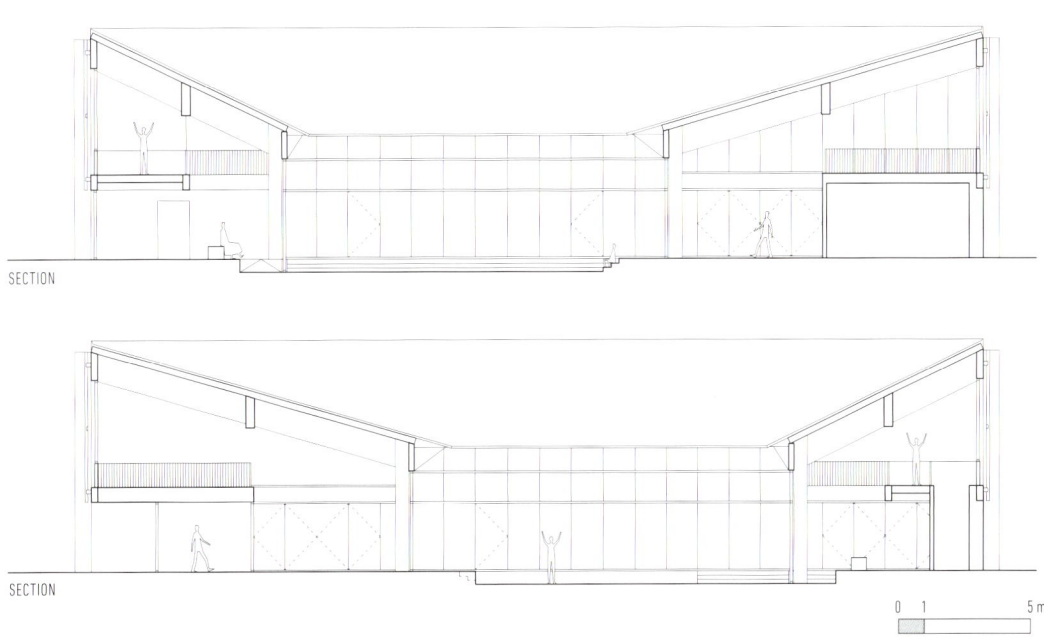

SECTION

SECTION

0 1   5 m

Planned Spontaneity

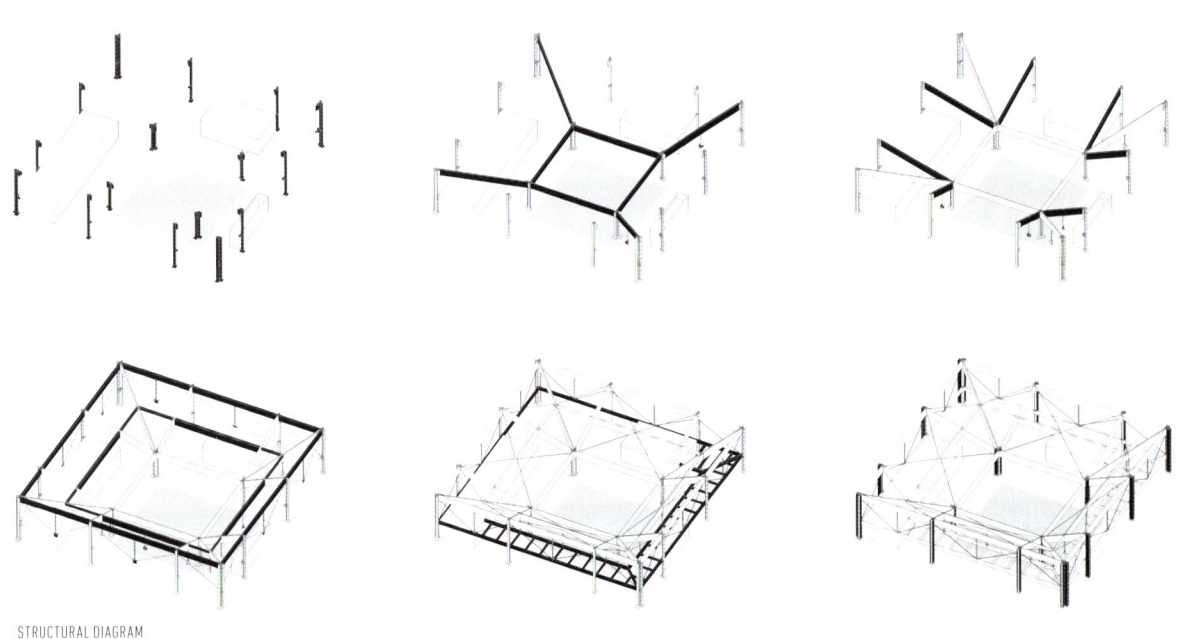

STRUCTURAL DIAGRAM

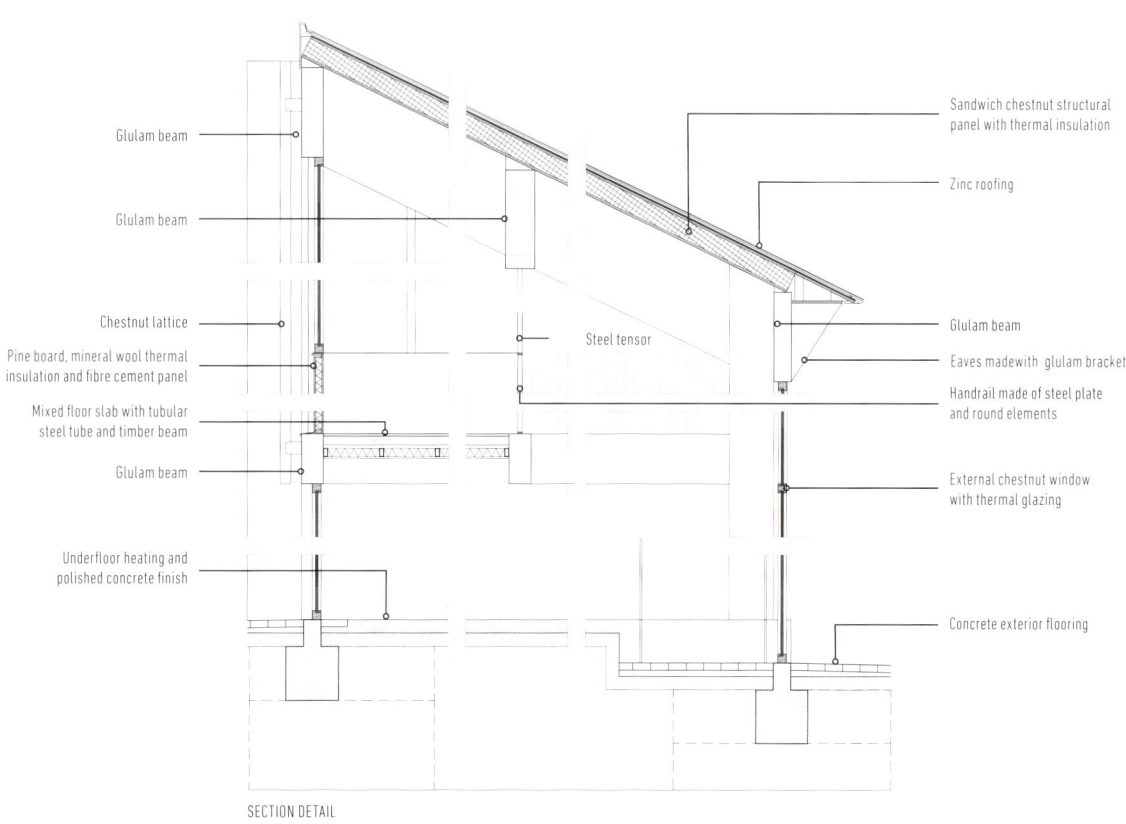

SECTION DETAIL

- Glulam beam
- Glulam beam
- Chestnut lattice
- Pine board, mineral wool thermal insulation and fibre cement panel
- Mixed floor slab with tubular steel tube and timber beam
- Glulam beam
- Underfloor heating and polished concrete finish
- Steel tensor
- Sandwich chestnut structural panel with thermal insulation
- Zinc roofing
- Glulam beam
- Eaves made with glulam brackets
- Handrail made of steel plate and round elements
- External chestnut window with thermal glazing
- Concrete exterior flooring

Planned Spontaneity   223

Communal Education

OGDEN CENTRE FOR FUNDAMENTAL PHYSICS AT DURHAM UNIVERSITY
INSTITUTIONAL
WOOD, CONCRETE, GLASS
DURHAM, UNITED KINGDOM
2017

The Ogden Centre for Fundamental Physics at Durham University is designed as continuous, stacked and interlocking forms. Clad in a ventilated timber rain screen built from responsibly sourced Scottish larch, the dynamic façade is punctuated with linear bands of operable strip windows and a series of outdoor terraces. In addition, curtain walls on the north and south faces bookend the form of the spiral and provide spectacular views of picturesque Durham city and Durham Cathedral.

Visitors to the Centre enter through a glazed lobby infused with light. The interiors employ a natural palette of soft gray concrete columns and ceilings, paired with warm wood finishes and frosted glass. Pushing the offices to the perimeter of the Centre allows each space to benefit from natural light and ventilation. Glazed doors and screens further transmit light from the exterior to the central atrium. Generous roof terraces create spaces for impromptu meetings or places to simply relax and enjoy fresh air. The entire program of the building is wrapped around a series of flexible, communal spaces.

The ground floor areas are open and available to the public. The second floor features a central social space where staff and research students can work together on an informal basis. The third floor houses additional office space and is open to the floors below.

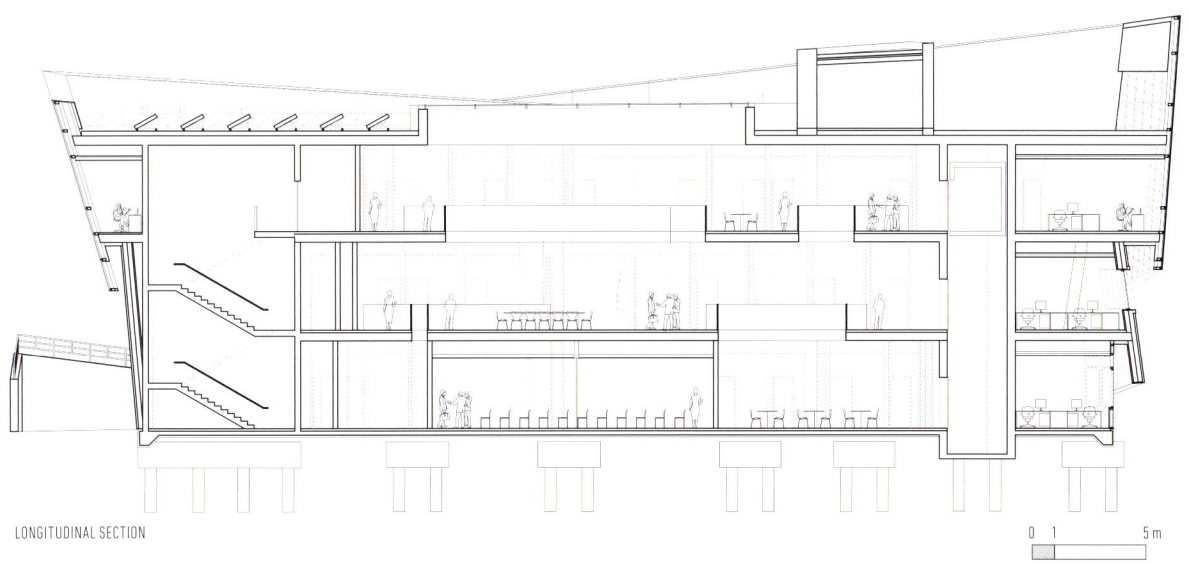

LONGITUDINAL SECTION

Communal Education 229

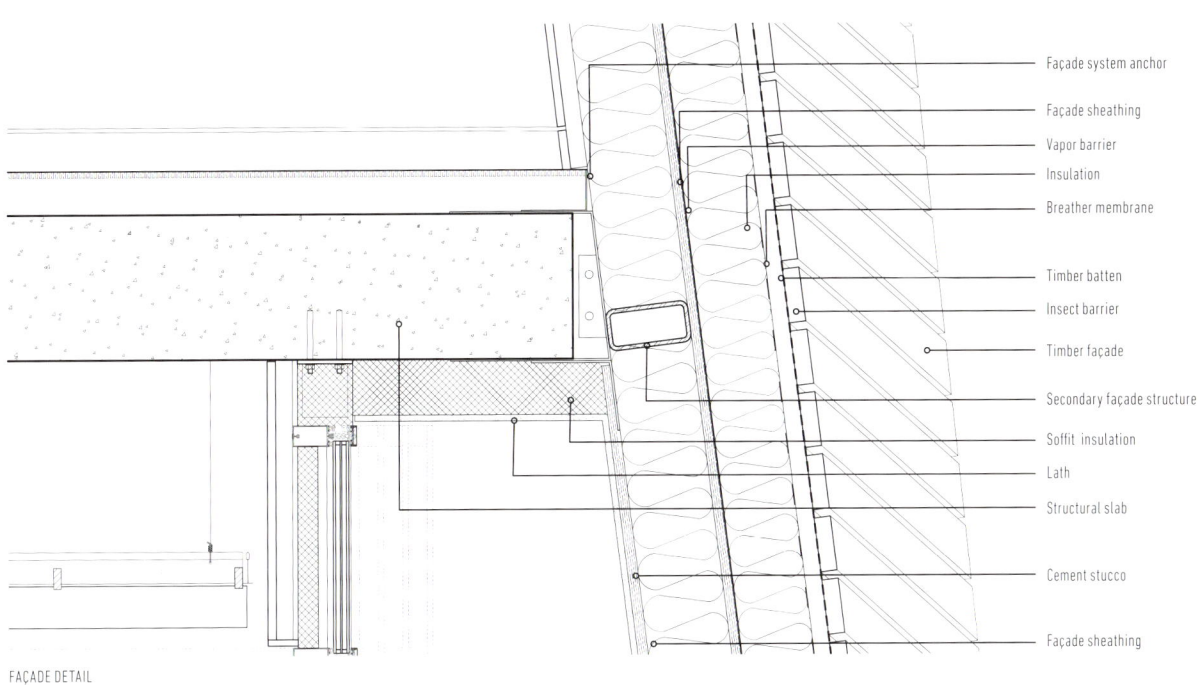

FAÇADE DETAIL

- Façade system anchor
- Façade sheathing
- Vapor barrier
- Insulation
- Breather membrane
- Timber batten
- Insect barrier
- Timber façade
- Secondary façade structure
- Soffit insulation
- Lath
- Structural slab
- Cement stucco
- Façade sheathing

Communal Education

Making It

BERLUTI MANUFACTURE
INDUSTRIAL
WOOD, STONE, STEEL, GLASS
FERRARA, ITALY
2016

South of Ferrara stands the new Berluti luxury shoemaking factory. The building's concept is derived from a dual logic: to make part of the site's industrial, technical, and infrastructural semiology disappear, and to endow the envelope with a vibrant and kinetic presence in order to diminish its mass. No technical protrusion or machinery are visible from outside.

The brand identity has been contextualized and transcribed in the architecture: wood is the dominant material of the building whose façades will develop a patina over time. Various sections, such as the battens in untreated red cedar, are rhythmically repeated across the lateral façades, which rise up from the ground to form a sun breaker, opening up the workshop to the landscape. By contrast, the main façade is smooth with an alternation of glazed panels and openable wooden panels (untreated red cedar) that liven up the façade and offer natural ventilation. Attention was given to sustainability, quality, and comfort of the workspaces, with an emphasis on natural ventilation and light. Natural materials were used in conjunction with a compact structure, photovoltaic panels, and an ability to control the energy consumption of the building.

Grandiose, with its grid of interlocking beams in resinous woods and projecting shadows, the agora connects all the spaces, and brings all crafts, skills, and know-how together at the heart of this place of manufacture.

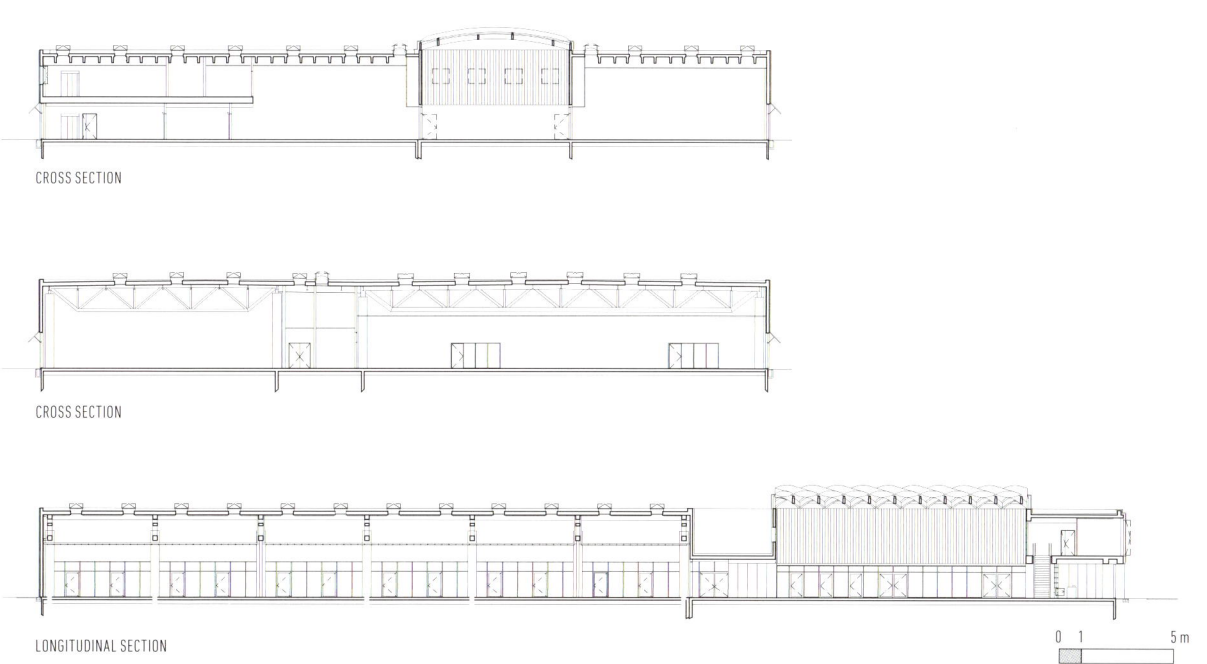

CROSS SECTION

CROSS SECTION

LONGITUDINAL SECTION

0 1 5 m

Making It 239

Making It   241

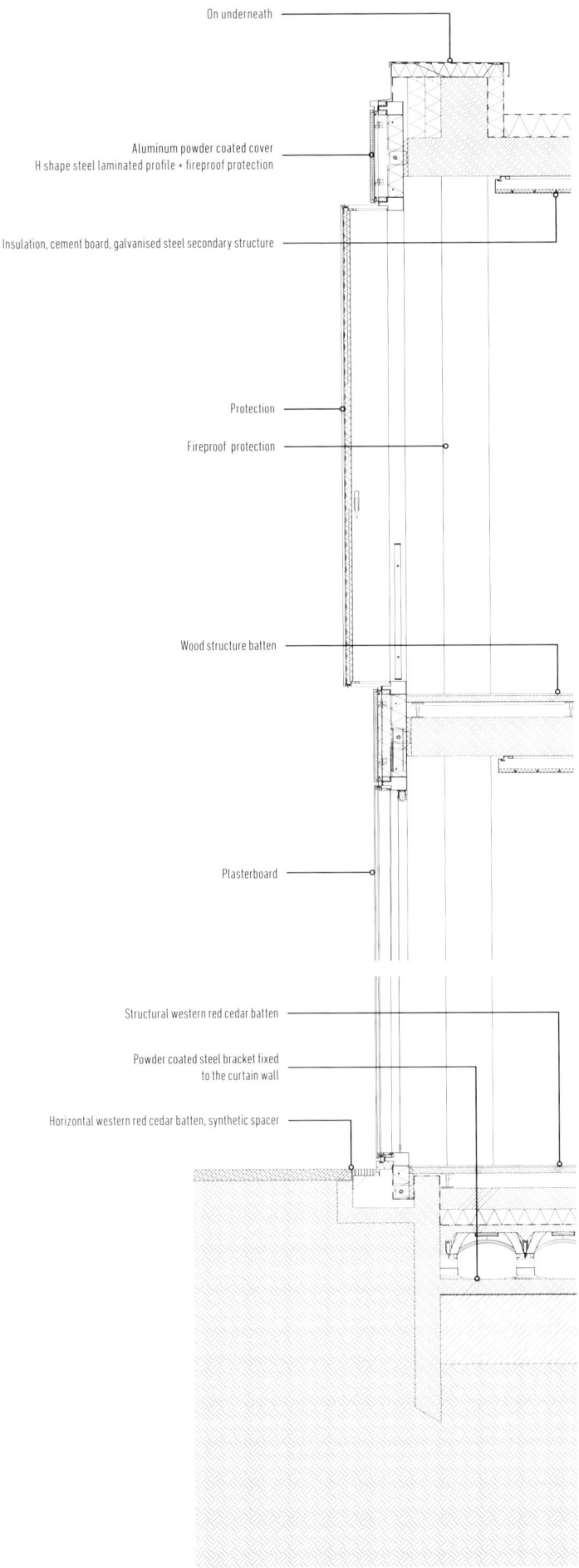

FAÇADE SECTION DETAIL

Worthwhile Delays

GIMNASIO MUNICIPAL DE SALAMANCA
CULTURAL
WOOD, STEEL, CONCRETE
SALAMANCA, CHILE
2016

Originally designed in 2007, the completion of this sport facility was delayed until 2016 for economic reasons. The delay allowed specific modifications based on observations and opportunities given by the site. The site itself housed an old gym and was next to a soccer field and an abandoned pool. The goal was to design a sports complex able to accommodate 2,000 spectators but the finished structure is also able to host public events and meetings.

The completed building is a pleasing combination of wood and steel. The north section receives the most sunlight and contains the gym, café, and administrative offices. The interior of the building is lit with indirect light. This light is achieved with a wooden roof placed upon the tribune concrete structure, organized through a game of interstitial spaces, the result of folded planes, like a paper plane.

The series of planes are located in relation to the sun trajectory. The lifted body has a faceted geometry related to the horizontal and diagonal ground-planes in which the building is inserted.

The project does not follow the structural optimum (of direct force transferences to the ground) but achieves a structural logic, a systematization of forcé transferences on different mismatches of structural elements. From wood to steel and from steel to concrete, forces finally arrive to the ground and foundations.

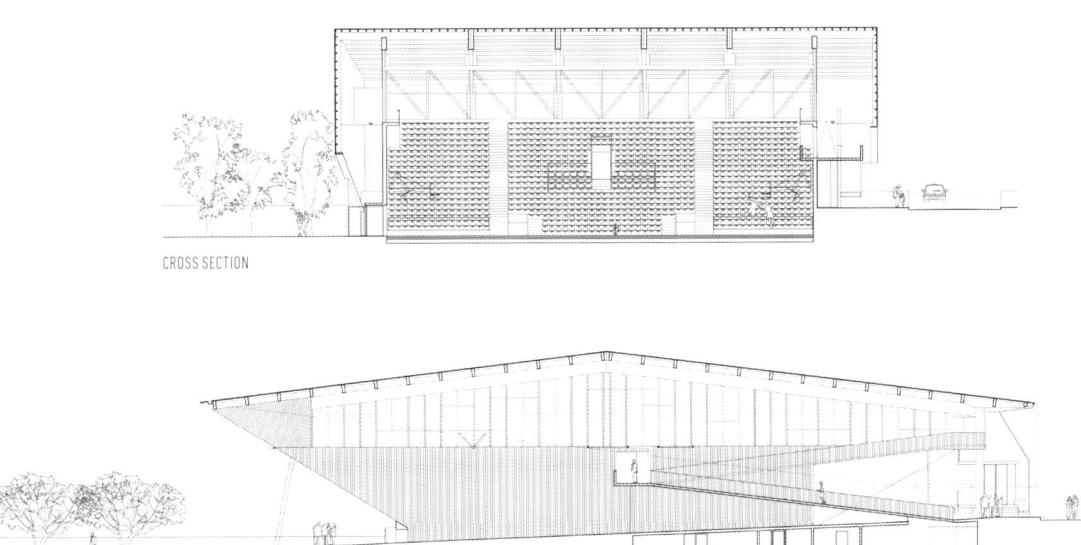

CROSS SECTION

LONGITUDINAL SECTION

Worthwhile Delays

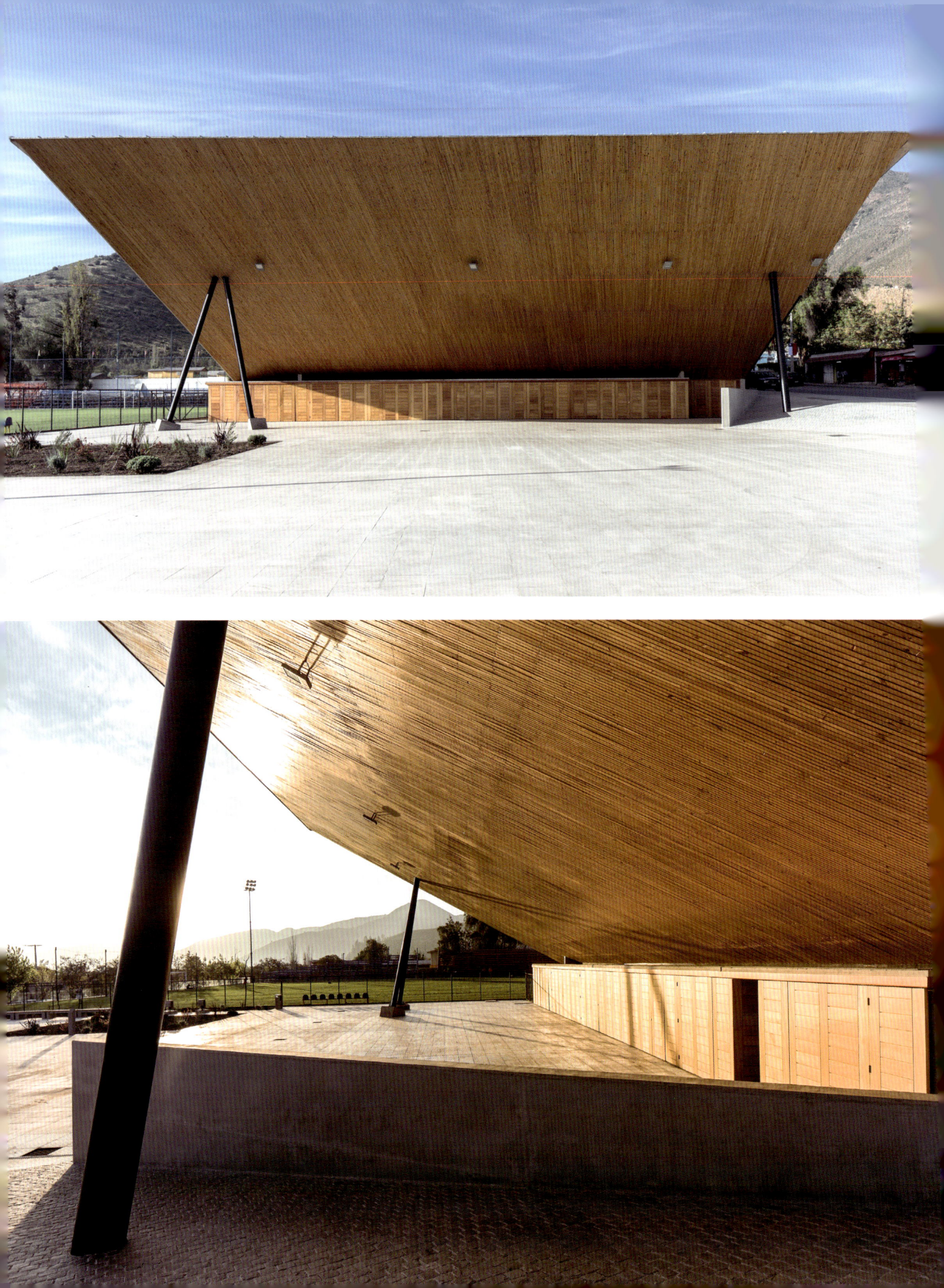

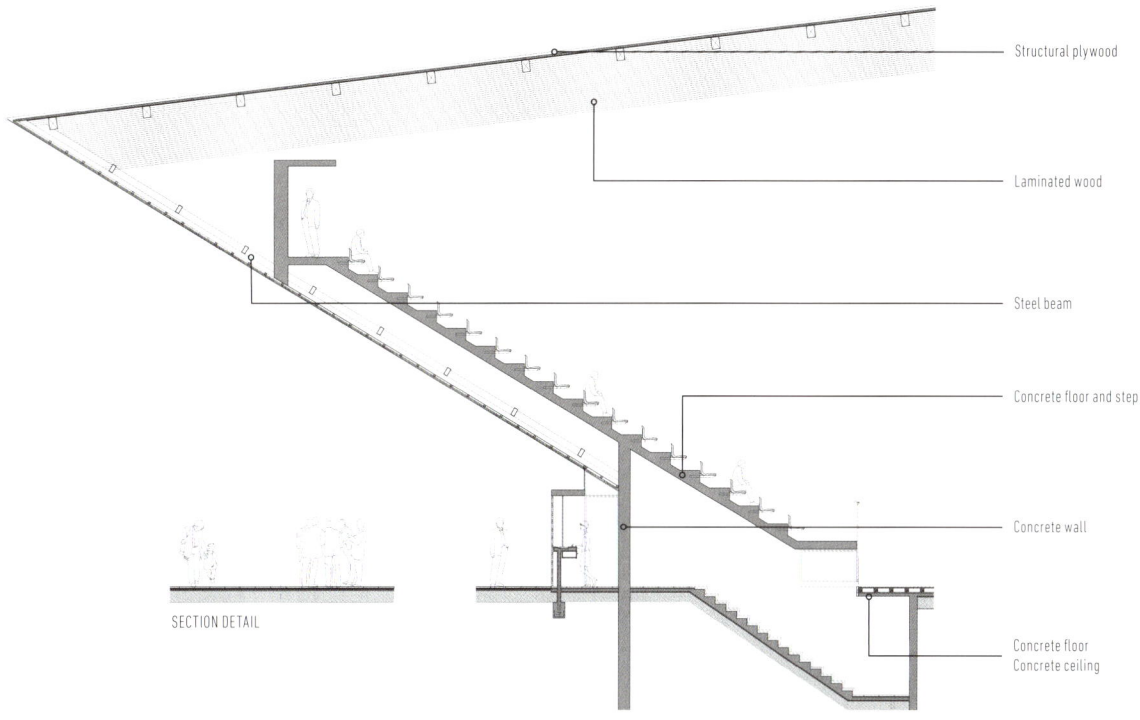

SECTION DETAIL

- Structural plywood
- Laminated wood
- Steel beam
- Concrete floor and step
- Concrete wall
- Concrete floor
- Concrete ceiling

Worthwhile Delays   **251**

Project Credits

**Weaving Together the
Strands of History** 14–21
Warwick Street
Squire and Partners
squireandpartners.com
Photography Gareth Gardner

Coming Together 22–27
Singkawang Cultural Center
PHL Architects
phlarchitects.com
Photography PHL Architects

Green and Open Space 28–35
Apartment in Binh Thanh
Sanuki Daisuke Architects
sanukiar.com
Photography Hiroyuki Oki

Letting the Light In 36–45
Fundación Santa Fe
El Equipo Mazzanti
elequipomazzanti.com
Photography Alejandro Arango

Brick in the Wall 46–55
Kent State Center for Architecture
and Environmental Design
Weiss/Manfredi
weissmanfredi.com
Photography Albert Vecerka of Esto

Private Spaces 56–63
KS Residence
Arquitetos Associados
arquitetosassociados.arq.br/
Photography Joana França

Reclaiming Acoustics 64–71
CKK Jordanki
Fernando Menis
menis.es/es/
Photography Jakub Certowicz

Historical Tradition 72–77
House for Solidarity
EllenaMehl Architects
lnamel.com
Photography Hervé Ellena

Echoing the Ocean 80–87
Atlantic Pavilion
Valdemar Coutinho Arquitectos
valdemarcoutinho.com
Photography João Morgado

Curtain Wall 88–95
Wenzhou Century Park Culture Club
Lacime Architects
lacime-sh.com
Photography Xingzhi Architecture

Luxury in Stone 96–103
Akelarre Hotel
mecanismo
mecanismo.org
Photography Kike Palacios

Elegant Lake Views 104–11
Bürgenstock Hotel
Rüssli Architekten AG
ruessli.ch
Photography Leonardo Finotti, Roger Frei

Pre-Hispanic Splendor 112–19
Huayacán
Alfredo Raymundo Cano, t3arc
t3arc.com
Photography Luis Gordoa

Looking at the Heroic Age 122–29
Museum of Troy
Yalin Architectural Studio
yalin-mimarlik.com
Photography Emre Dörter

Blending In 130–39
New Entrance Hall of Museum of the Middle Ages
Bernard Desmoulin Architecte
desmoulin-architectures.com
Photography Michel Denance, Célia Uhalde

Revisiting Ancient Tradition 140–45
Shui Cultural Center
West-Line Studio
china-west-line.com
Photography Haobo Wei, Jingsong Xie

A-Framing the View 146–53
Sierra
BOSS.architecture
bossarch.com
Photography James Florio

Private Access 156–61
House Behind the Roof
Superhelix Pracownia Projektowa –
Bartlomiej Drabik
superhelix.pl
Photography Bartlomiej Drabik

New Façades 162–71
Ibiza Gran Hotel
Colmenares Vilata Arquitectos
colmenaresvilata.com
Photography Imagen Subliminal
(Miguel de Guzmán + Rocío Romero)

The Heart in Ikast 172–81
Hjertet
C.F. Møller Architects
cfmoller.com
Photography Adam Mørk, Thomas Mølvig,
Thomas Olsen, Julian Weyer

Stacking the Terraces 182–87
Inter Crop Office
Stu/D/O Architects
stu-d-o.com
Photography Stu/D/O, Chaovalith Poonphol

Rise Like a Phoenix 188–93
Polyvalent Hall
LOCALARCHITECTURE
localarchitecture.ch
Photography Matthieu Gafsou

Treasure Box 194–201
Steinhardt Museum of Natural History
Kimmel Eshkolot Architects
kimmel.co.il
Photography Amit Geron

Student Rhythm 202–7
Arkansas Bear Claw
modus studio
modusstudio.com
Photography Timothy Hursley, Kiara Luers

Project Credits continued

Disappearing Trick 208–15
Holmen Aquatics Center
ARKÍS arkitektar
ark.is
Photography Tove Lauluten, Lasse Leonhardsen

Planned Spontaneity 216–25
Impluvium Community Center
RAW/deAbajoGarcia Oficina de Arquitectura
deabajogarcia.com
Photography Montse Zamorano

Communal Education 222–35
Ogden Centre for Fundamental Physics at Durham University
Studio Libeskind
libeskind.com
Photography Hufton+Crow

Make It 236–43
Berluti Manufacture
Barthélémy Griño Architectes, Paris
barthelemygrino.com
Photography Arnaud Schelstraete

Worthwhile Delays 244–53
Gimnasio Municipal de Salamanca
CARREÑO SARTORI ARQUITECTOS
carrenosartori.com
Photography Marcos Mendizabal

All photography is attributed in the Project Credits unless otherwise noted below.

Pages 1, 5, 78–79, 248: Xurzon/iStock
Pages 2–3: Michel Denance, Célia Uhalde (Bernard Desmoulin Architecte, New Entrance Hall of Museum of the Middle Ages)
Pages 5, 12–13: themacx/iStock
Pages 5, 120–21: photostio/iStock
Pages 5, 154–55: sankal/iStock

Endpapers: Jakub Certowicz (Fernando Menis, CKK Jordanki)

Front cover image: Michel Denance, Célia Uhalde (Bernard Desmoulin Architecte, New Entrance Hall of Museum of the Middle Ages)
Back cover image: Hiroyuki Oki (Sanuki Daisuke Architects, Apartment in Binh Thanh)
Back cover inset image (behind text): Xurzon/iStock